Contents

1. Properties of a Transiting Extrasolar Planet 1
2. Temperature Scales 9
3. Measuring Angles 13
4. Motion on the Celestial Sphere 23
5. Motions of the Moon 27
6. Lunacy: Moon Phase Exercise 31
7. The Phases of Venus 33
8. Mapping the Solar System from Earth 37
9. Moon Phases 43
10. The Seasons and the Sun's Path on the Celestial Sphere 53
11. Force, Acceleration, and Gravity 55
12. Kepler's Laws and Elliptical Orbits 59
13. How Long are the Days on Mercury and Venus? 63
14. Electromagnetic Radiation and Thermal Spectra 67
15. A Scale Model of the Solar System 69
16. Kuiper Belt Objects 89
17. Age Dating Through Radioactive Decay 103
18. Tidal Forces and the Roche Limit 105
19. Stellar Masses with Newton's Version of Kepler's Third Law 109
20. Luminosity, Brightness, and the Inverse Square Law 111
21. Elementary Particles and Forces Review 113
22. Fusion in the Sun: the Proton-Proton Chain 115
23. Spectra 117
24. The Diameters And Luminosities of Stars 119
25. Stellar Classification 139
26. The Hertzsprung-Russell (H-R) Diagram 141
27. Stellar Evolution 145

28. Special Relativity: Time Dilation and Length Contraction 151
29. Worldlines and Light Travel Time 157
30. What Lurks at the Center of the Milky Way? 159
31. Quasar Jets: a Superluminal Optical Illusion 161
32. The Expansion of Space 163
33. The Acceleration of the Milky Way 167
34. Civilizations in Our Galaxy: the Drake Equation 169

Properties of a Transiting Extrasolar Planet

NAME ______________________ ID# ______________________

DATE ______________________ LAB SECTION# ______________________

Write your answers on each page of this lab and hand the entire lab in.

In this lab you will study a transiting extrasolar planet in a circular orbit around a main sequence star. You will use the star's temperature, luminosity, and motion and the lightcurve of one transit of the planet to learn as much as you can about the planet's physical parameters. Along the way, you will compare your planet to planets in our solar system.

Below we have illustrated what happens during a transit: a planet (small circle) moves between us and the star it orbits (big circle). As time goes on after a transit starts, the lightcurve of the star (bottom) shows that an increasing amount of the star's light is blocked until a maximum depth is reached (time t_2, Second Contact). The planet begins moving out from in front of the star at time t_3 (Third Contact), and the transit ends at time t_4.

1. Your lab coordinator will give you a unique chart of a simulated lightcurve of one transit of a unique planet and star. Near the bottom of the chart, the planet is given a number.

   Record the number of your planet here: __________

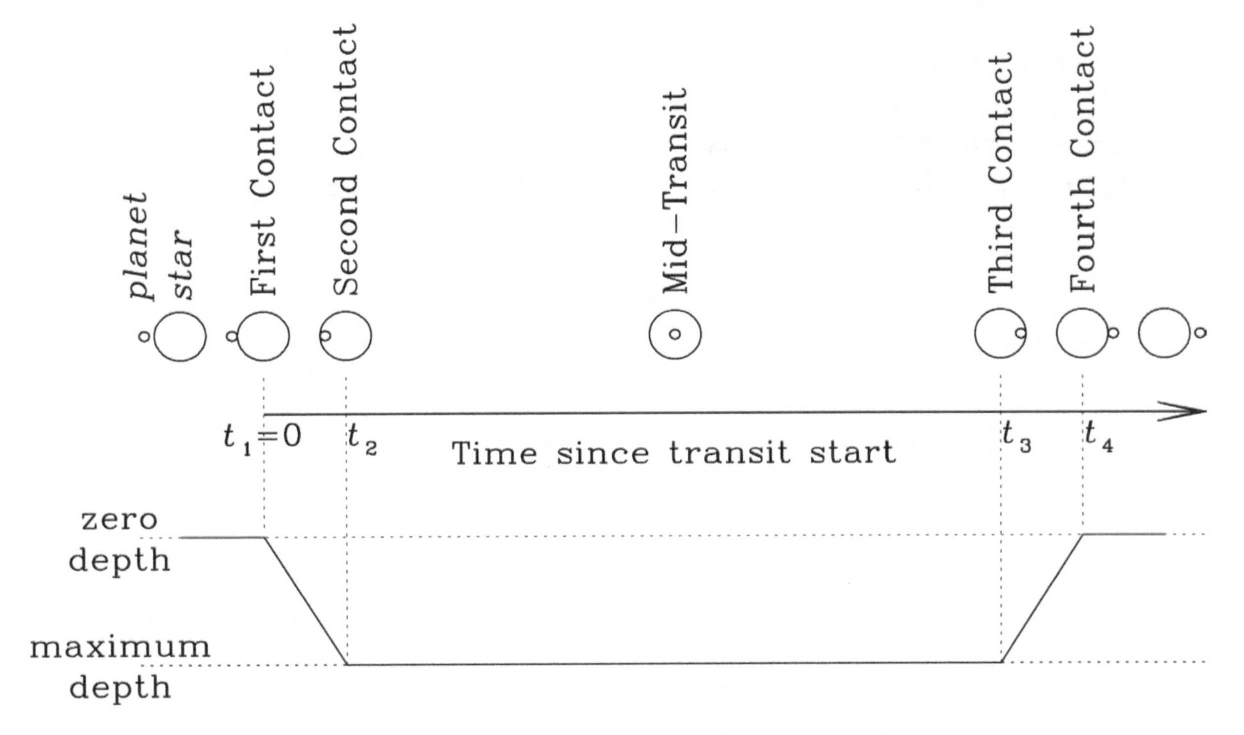

Figure 1: A planet (small circle) transiting across the exact center of its star (large circle).

Outline

Here is the information you will determine about your planet and star in this lab:

- Find the luminosity, mass, and radius of your star from its temperature.
- Find the radius of your planet from its maximum transit depth.
- Find the distance between your planet and star using Newton's version of Kepler's 3rd Law.
- Find what part of your star your planet transited from the time it took to transit.
- Find your planet's mass from your star's Doppler motion.
- Find your planet's gravity and density from its mass and radius.
- Find the cloud-top temperature on your planet (extra credit).
- Summarize how your planet compares to planets in our solar system.

From Temperature to Luminosity, Mass, and Radius of Your Star

Your chart includes the temperature of your star in Kelvin. Because your star is a main sequence star, its temperature is uniquely related to its radius and mass. Using Table 1, look up the luminosity *L*Star of your star relative to the Sun. (For example, a star with $T_{Star} = 10,000$ K has $L_{Star} = 56$, meaning that it emits 56 times more energy than the Sun.)

Table 1: Luminosities and Masses of Main Sequence Stars Relative to the Sun

Star's Temperature T_{Star}	4000 K	5000 K	5500 K	6000 K	8000 K	10000 K
Star's Luminosity L_{Star}	0.11	0.28	0.80	1.4	11	56
Star's Mass M_{Star}	0.54	0.75	0.90	1.1	2.0	2.9

We can use your star's luminosity and temperature to find its radius relative to the Sun's radius. We know that luminosity is proportional to area times temperature[4], that the area of a star is 4π times radius², and that the Sun has $L_{Sun} = 1$ and $T_{Sun} = 5780$ K. The ratio of your star's luminosity to the Sun's luminosity can therefore be used to get your star's radius relative to the Sun's:

$$\frac{L_{Star}}{L_{Sun}} = \frac{L_{Star}}{1} = \frac{4\pi(\text{Star's Radius})^2 T_{Star}^4}{4\pi(\text{Sun's Radius})^2(5780\,\text{k})^4} \rightarrow \frac{\text{Star's Radius}}{\text{Sun's Radius}} = \sqrt{L_{Star}} \Big/ \left(\frac{T_{Star}}{5780\,\text{K}}\right)^2$$

2. Take the square root of L_{Star} and call it A.

$$A = \sqrt{L_{Star}} = \_\_\_\_\_\_ \qquad \text{(A)}$$

3. Divide T_{Star} by 5780, square the result, and call it B.

$$B = \left(\frac{T_{Star}}{5780}\right)^2 = \_\_\_\_\_\_ \qquad \text{(B)}$$

4. Now divide A by B to get R_{Star}, your Star's radius in terms of the Sun's radius:

$$R_{Star} = A/B = \_\_\_\_\_\_ \qquad \text{(C)}$$

Radius of Your Planet

The maximum depth of a planet's transit gives the planet's area as a fraction of the star's area. Because the area of a circle is π(Radius)^2, the radius of the planet as a fraction of the star's radius is the square root of the depth:

$$\text{Depth} = \frac{\text{Planet Area}}{\text{Star Area}} = \frac{\pi(\text{Planet Radius})^2}{\pi(\text{Star Radius})^2} \rightarrow \frac{\text{Planet Radius}}{\text{Star Radius}} = \sqrt{\text{Depth}}$$

5. Measure the maximum depth of your planet's transit on the chart you were given. Note that the numbers increase DOWNWARDS on the graph. You should measure the depth to within at least 0.0001 (at least four numbers to the right of the decimal point, but no more than five).

Maximum Depth = _._ _ _ _ _ (D)

6. Take the square root of the depth to find your Planet's radius relative to your Star's radius (call it E). Write down five numbers to the right of the decimal point.

$\sqrt{\text{Maximum Depth}}$ = _._ _ _ _ _ (E)

7. It's easier to understand the planet's size relative to Earth's size. The Sun has a radius equal to 109 times the Earth's radius. Therefore we just have to multiply by 109 times the size of your Star relative to the Sun (C) to find the planet's radius relative to the Earth's radius (F):

Planet's Radius relative to Earth's Radius = $109 \times C \times E$ = _ _._ _ (F)

(In your answer you should have only one or two numbers to the *left* of the decimal point and should keep two numbers to the right of the decimal point.)

8. Use Table 2 to answer the following question comparing your planet's radius to the radii of some planets in our Solar System.

My planet's radius is in between that of the planets ________________ and ________________. (For example, if your planet's radius relative to Earth's was 0.7, the correct answer is "between Mercury and Venus", because those are the planets with radii closest to yours.)

Table 2: Some Useful Properties of Some Planets in our Solar System

Planet	Radius/ Earth's	Mass/ Earth's	Density in grams/cm^3	Semi-major Axis / Star's Radius	Gravity / Earth's	Cloud-top Temperature
Mercury	0.38	0.055	5.43	83.3	0.38	449 K (176 C)
Venus	0.95	0.815	5.25	155.6	0.90	329 K (56 C)
Earth	1.00	1	5.52	215.2	1	279 K (6 C)
Jupiter	11.19	317.9	1.33	1120.0	2.54	123 K (–150 C)
Saturn	9.46	95.18	0.70	2053.0	1.06	91 K (–183 C)
Neptune	3.81	17.13	1.64	6469.0	1.18	51 K (–222 C)

Distance of Your Planet from its Star

Newton's version of Kepler's Third Law states that for planets in circular orbits around a star of mass M_{Star}, the planet's orbital radius a and its orbital period P are related by

$$a^3 \propto M_{Star} \times P^2$$

as long as the mass of the planets is much less than M_{Star}. Specifically, if a is measured relative to our own Sun's radius and P is in Earth days, then

$$\left(\frac{a}{R_{Sun}}\right)^3 = 74.7 \times M_{Star} \times (P_{days})^2$$

9. Look at the bottom of your individual chart and find the number of days between each transit; that is the period P_{days} of your planet's orbit, measured in Earth days. Square P_{days}, multiply the result by 74.7 times M_{Star} (from Table 1), and call the result G.

$$G = 74.7 \times M_{Star} \times \left(P_{days}\right)^2 = \_\_\_\_\_\_\_\_\_\_ \qquad \text{(G)}$$

10. Take the cube root of G and divide it by the size of your star (C) to get a, your planet's orbital radius in units of the star's radius. (Taking a cube root is the same as raising to the power of 0.33.)

$$a = \frac{\sqrt[3]{G}}{C} = \_\_\_\_\_\_ = \_\_\_\_\_\_ \qquad \text{(H)}$$

11. Which solar system planet in Table 2 has an orbital radius in units of its star's radius which is most similar to your planet's?

What Part of Your Star Did Your Planet Cross?

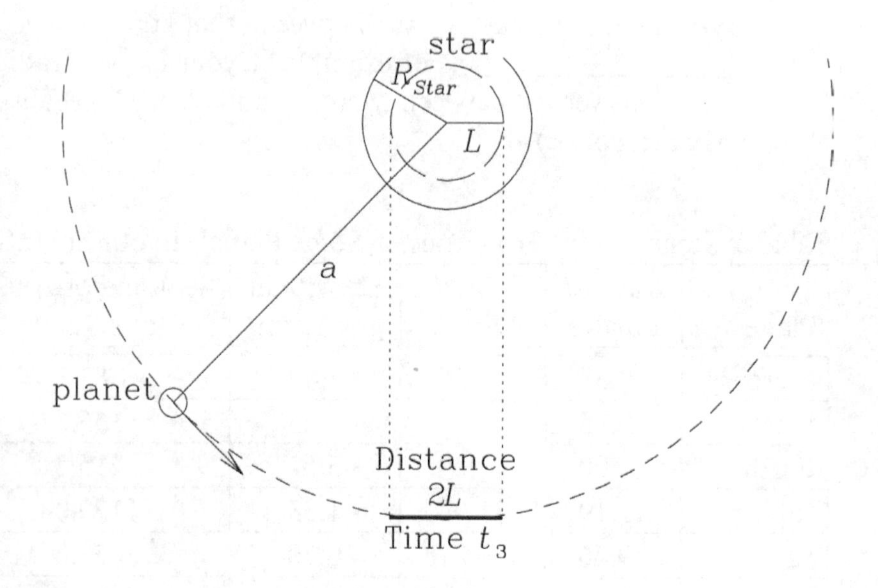

Figure 2: A planet orbiting its star, looking top-down onto the orbit. Earth is in the direction of the bottom of the page. A transit occurs when the planet is between the star and Earth. The star has diameter $2R_{Star}$ and the planet travels distance $2L$ during the transit. If the planet transits across the center of its star, $2L=2R_{Star}$, and the duration of the transit, t_3, is the maximum possible. If a planet transits off-center, $2L<2R_{Star}$, and the transit is shorter.

The distance around a circular orbit is 2π times the radius of the orbit. So your planet travels a distance $2\pi a$ in time P_{days} (one orbital period), which means its velocity is $v=2\pi a/P_{\text{days}}$. Your planet also travels a distance $2L$ in time t_3 (and distance $2L + 2R_{\text{Planet}}$ in time t_4), which means its velocity is $v=2L/t_3$. Those two expressions for the velocity must equal each other, which happens if:

$$\text{Fraction of star's diameter crossed during transit} = 2\text{L} = \frac{0.131 \times a \times t_3}{P_{\text{days}}}$$

where t_3 is measured in hours and a is measured in units of the star's radius.

12. Measure t_3, the time of Third Contact (end of maximum depth) on your transit chart (bottom panel of your handout), in hours. Multiply t_3 by 0.131 and by your planet's orbital radius in units of the star's radius:

$$J = t_3 \times 0.131 \times a = \_\_\_\_\_\_ \times 0.131 \times \_\_\_\_\_\_ = \_\_\_\_\_\_ \qquad \text{(J)}$$

13. Look at the bottom of your individual chart and find the number of days between each transit; that is the period P_{days} of your planet's orbit, measured in Earth days. Divide J by P_{days} to get $2L$, the fraction of your star's diameter crossed by your planet during one transit:

$$2L = \frac{J}{P_{\text{days}}} = \_\_\_\_\_\_ = \_\_\_\_\_\_ \qquad \text{(L)}$$

Figure 3 shows half of your star. The straight lines across it show paths taken by transiting planets. Each path crosses the fraction of the star's diameter indicated on the left-hand axis.

14. Draw a straight line on Figure 3 indicating the path of your transiting planet across the face of your star, using the left- and right-hand axis labels as a guide. For example, if you found a value of $2L$=0.990, you would draw a straight line between lines b and c, but closer to c.

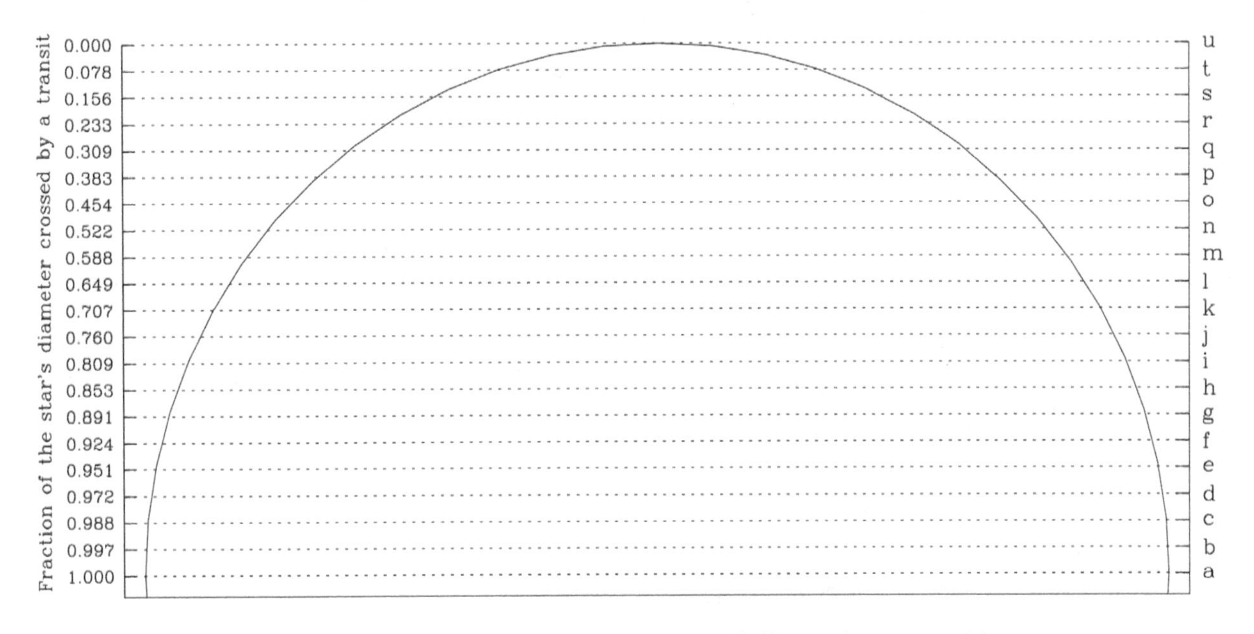

Figure 3: The lines show some of the paths planets can follow when transiting your star.

Mass of Your Planet

To measure the mass of your planet requires measuring the motion of the star around which it orbits. A planet in a circular orbit moves in a circle around the center of mass of the star- planet system, at the same time as the star moves in a much smaller circle around the same point. The planet and star are on exactly opposite sides of the center of mass at all times, which means that the star and planet velocities are related by:

$$v_p = v_s \, M_{\text{Star}} / M_p$$

We also know that the planet travels a distance 2πa in one orbital period P_y (measured in Earth years), so the planet's velocity also equals:

$$v_p = 2\pi a / P_y$$

Those two equations for v_p must equal each other, and if we combine them we get an expression for the planet's mass relative to the star's mass, which we will use below:

$$M_p / M_{\text{Star}} = v_s P_y / 2\pi a$$

The only complication is that we measure P in days, not years, which introduces another number we will combine with the factor of 2π in our calculations.

15. Measure v_s, the **maximum** radial velocity of the star in the top panel of your handout. Multiply that number by P_{days}, the period of your planet's orbit in Earth days (see Question 9). Then divide by H (your planet's orbital radius in units of the star's radius) times C (your star's radius in units of the Sun's radius). Call the result N.

$$N = \frac{v_s \times P_{\text{days}}}{H \times C} = \frac{\quad \times \quad}{\quad \times \quad} = \_\_\_\_\_\_\_\_\_\_ = \_\_\_\_\_\_\_\_\_\_ \qquad \text{(N)}$$

16. Divide N by 50,578 and call the result O; that is your planet's mass relative to its star.

$$O = \frac{N}{50{,}578} = \frac{\textit{Planet Mass}}{\textit{Star Mass}} = \_.\_\_\_\_\_\_ \qquad \text{(O)}$$

17. It is easier to understand a planet's mass if it is in Earth masses. The Sun has a mass of 332,746 Earth masses, so your star has a mass of (332, 746 × M_{Star}) Earth masses. You can multiply O by 332,746 × M_{Star} to get M, the mass of your planet relative to Earth's mass.

$$M = O \times 332{,}746 \times M_{\text{Star}} = \_\_\_\_\_ \times 332{,}746 \times \_\_\_\_\_ = \_\_\_.\_\_ \qquad \text{(M)}$$

18. Use Table 2 to answer the following question.

My planet's mass is in between that of the planets ____________ and ____________.

Density of Your Planet

Now that you know your planet's radius (size) and mass, you can determine its density (mass divided by volume) and compare it to high-density terrestrial planets and low-density giant planets in our solar system.

19. Cube your planet's radius relative to Earth (F) and call the result Q:

$$F^3 = F \times F \times F = Q = \_\_\_\_\_\_\_\_ \qquad \text{(Q)}$$

20. Multiply your planet's mass M by 5.5 and divide it by Q to get your planet's density R in grams per cubic centimeter (cm^3):

$$\frac{5.5 \times M}{Q} = R = \_\_\_\_\_\_ \qquad \text{(R)}$$

21. According to Table 2, your planet's density R is most similar to the density of (circle one):
    - A terrestrial planet like Mercury, Venus, or Earth
    - A gas giant planet like Jupiter
    - A low-density gas giant planet like Saturn
    - An ice giant planet like Neptune

How Much More Would You Weigh on Your Planet?

What you feel as weight is the force of Earth's gravity pulling on your body. The more mass you have, the more Earth pulls on you. The force of any planet's gravity per unit mass (one g) is proportional to the planet's mass divided by its radius squared: $g \propto M/R^2$. If you were to stand on another planet (or on a blimp in its outer atmosphere, if it doesn't have a solid surface), you would feel a different weight. Let's see how much more you would weigh on your planet.

22. Square your planet's radius relative to Earth (F) and call the result S:

$$F^2 = F \times F = S = \_\_\_\_\_\_\_\_ \qquad \text{(S)}$$

23. Divide your planet's mass M by S; the result—call it W—is your planet's surface gravity relative to Earth's:

$$\frac{M}{S} = W = \_\_\_\_\_\_\_\_ \qquad \text{(W)}$$

In other words, on your planet you would weigh W times as much as on Earth.

24. Use Table 2 to compare your planet's surface gravity to that of planets in our solar system:

    The surface gravity of my planet is greater than that of __________ but less than that of __________. (If there is no planet in Table 2 with a larger or smaller surface gravity than your planet's, put 'no planet in Table 2' in the appropriate blank.)

Summary of Your Planet

25. In one sentence, give your planet's properties as compared to planets in our solar system. For example: "My planet orbits at the distance of Venus but is bigger than Neptune, with a mass close to Jupiter's, giving it a terrestrial planet's density and a surface gravity four times Earth's."

____

____

____

Extra Credit

Cloud-Top Temperature of Your Planet

The cloud-top temperature of a fast-rotating planet is determined by the balance between how much heat the planet absorbs from the star and how much heat it emits into space. The size of the planet does not matter: a bigger planet absorbs more heat but also emits more heat. What does matter is how close the planet is to the Sun, and how much heat from the Sun it reflects rather than absorbs.

If we assume that the planet absorbs all the heat that hits it, we can estimate the temperature at the top of its atmosphere. (If the planet has no atmosphere, the cloud-top temperature will just be the surface temperature. But if the planet does have an atmosphere, the greenhouse effect can increase the surface temperature above the cloud-top temperature.)

The cloud-top temperature of a fast-rotating planet orbiting a star of temperature T_{Star} is

$$\text{Cloud-top Temperature} = \frac{0.7071 \times T_{\text{Star}}}{\sqrt{\text{Planet's orbital radius a, relative to the star's radius}}}$$

26. **[Extra Credit]** Calculate your planet's cloud-top temperature in Kelvin:

$$T = \frac{0.7071 \times T_{\text{Star}}}{\sqrt{H}} = \_\_\_\_\_\_\_\_ \text{K} \qquad \text{(T)}$$

27. **[Extra Credit]** Convert your planet's cloud-top temperature to degrees Celsius:

$$U = T - 273.15 = \_\_\_\_\_\_\_\_ \text{C} \qquad \text{(U)}$$

28. **[Extra Credit]** Use Table 2 to compare your planet's cloud-top temperature to that of planets in our solar system. (If there is no planet in Table 2 with a lower or higher cloud-top temperature than your planet's, put 'no planet in Table 2' in the appropriate blank.)

The cloud-top temperature of my planet is greater than that of __________ but less than that of __________.

Unit 1.3 Temperature Scales

Objective

To become acquainted with the relations between the three most common temperature scales: Celsius, Fahrenheit and Kelvin

Introduction

Temperature is a measure of the energy of the particles which make up a body or an environment, like a cup of water or the atmosphere of a star, or rock on the surface of a planet. The temperature is related to the average speed of atoms or molecules which make up an object. A hot object, like the water molecules in a cloud of steam, are moving rapidly and chaotically, with each molecule containing relatively large amounts of energy. By contrast, the low-velocity atoms of water ice are held stationary in a solid crystal structure. As the ice is heated, the velocity of the molecules begins to break apart those bonds, reducing the crystal structure to a liquid puddle and eventually an expanding gas cloud.

Three temperature scales are most commonly used in everyday life, science, and industry.

The **degree Celsius (°C)** temperature scale was devised by the Swedish astronomer Anders Celsius in 1742. This scale is based on the behavior of pure water, which freezes at 0 °C and boils at 100 °C under standard atmospheric conditions (at sea level on the Earth). Therefore, there are 100 degrees between these points. This scale is used throughout the science and almost in all countries of the world except the United States.

The **Kelvin (K)** temperature scale, named after the British Lord Kelvin (William Thomson, Baron Kelvin of Largs) is an extension of the degree Celsius scale down to *absolute zero or 0 K,* a hypothetical temperature at which all atomic and molecular motions cease. *Absolute zero* is the lowest temperature hypothetically possible at which no heat exists. On the Kelvin temperature scale, water freezes at 273 K (more precise value: 273.16 K) and boils at 373 K. Absolute zero (0 K) or –273 °C is the starting point for the Kelvin scale. Since nothing can be colder than 0 K, there are no negative temperatures on the Kelvin scale.

The step size of the Kelvin and the Celsius temperature scales is same, because water must be heated by 100 K or 100 °C to go from its freezing to melting point. Scientists throughout the world (including the United States) prefer the Kelvin scale because it is closely related to the physical meaning of the temperature.

Note: Temperatures on this scale are called "kelvin," *not* degrees kelvin. Further, kelvin is *not* capitalized, and the symbol (capital K) stands alone with no degree symbol.

The **degree Fahrenheit (°F)** temperature scale, now antiquated, is still used by many in the United States. The German physicist Gabriel Fahrenheit introduced this scale in the early 1700s and he intended 0 °F to represent the coldest temperature achievable at that time and 100 °F to represent the temperature of a healthy human body. As you might know, normal body temperature is closer to 98.6 °F, suggesting that when he conducted his experiment, either he was having a fever, or his thermometer was inaccurate. Lastly, it is believed that he might have used a cow's temperature instead of his own. On this scale, water freezes at 32 °F and boils at 212 °F. Therefore, there are 180 degrees between these points. The step size of the Fahrenheit degree is smaller than that of the Celsius degree (or 1 kelvin). In other words, a degree Celsius (or a kelvin) is 180/100 (which is 9/5 or 1.8 times) the size of the degree Fahrenheit. An increase of 1 kelvin is equivalent to a Fahrenheit temperature increase of nearly 2 degrees. Notice that the degree Fahrenheit is a non-metric temperature scale, while the degree Celsius and the Kelvin temperature scales are metric scales (based on multiples of 10).

Note that the United States is the only country that uses Fahrenheit temperatures for shelter-level (surface) weather observations. However, since July 1996 all surface temperature observations in the National Weather Service METAR/TAF reports are transmitted in degrees Celsius.

Equations and Constants

Fahrenheit to Celsius Conversion $$C = \frac{(F-32)}{1.8}$$

Fahrenheit to Kelvin Conversion $$K = \frac{(F-32)}{1.8} + 273.16$$

Celsius to Fahrenheit Conversion $$F = (C \times 1.8) + 32$$

Celsius to Kelvin Conversion $$K = C + 273.16$$

Kelvin to Fahrenheit Conversion $$F = (K - 273.16) \times 1.8 + 32$$

Kelvin to Celsius Conversion $$C = K - 273.16$$

NAME ______________________________ ID ______________________________

DUE DATE __________ LAB INSTRUCTOR ___________ SECTION __________

Worksheet # 1

Answer the following questions related to the temperature scale

1. Which temperature scale or scales begin at zero?

2. Which temperature scale or scales allow for negative temperatures?

3. At what temperature does water freeze:

   On the Fahrenheit scale:

   On the Celsius scale:

   On the Kelvin scale:

4. A 1 degree temperature change on the Fahrenheit scale is equal to how many degree of temperature change on the Celsius scale?

5. Normal human body temperature is 98.6 °F; what is the healthy human body temperature:

   On the Celsius scale:

   On the Kelvin scale:

6. The color of a hot metal is directly related to the temperature of the metal. The coil on a stovetop burner will begin to glow a dim, deep red at 390 °C. At what temperature does a stove coil begin to glow:

   On the Celsius scale:

   On the Kelvin scale:

Continue....

7. The boiling point of oxygen – where oxygen transitions from a liquid puddle into a gas cloud – occurs at –183 °C. At what temperature does that process happen:

   On the Fahrenheit scale:

   On the Kelvin scale:

8. When temperatures drop below 63.15 K, nitrogen freezes, turning from liquid nitrogen into solid ice. At temperatures above 77.36 K, liquid nitrogen turns into a gas. On the dwarf planet Pluto, the wintertime temperature is low enough to freeze the nitrogen atmosphere, turning the atmosphere into a gentle snow of ice crystals. In the summertime, the temperature is high enough to turn nitrogen into a gas. If a thermometer on the surface of Pluto reads –214.15 °C, show whether Pluto's nitrogen will be frozen into ice, existing in puddles, or in a gaseous form making up an atmosphere.

9. The Sun's surface temperature is 5770 K. What is the temperature of the Sun:

   In Fahrenheit:

   In Celsius:

10. Absolute 0 on the Kelvin scale is 0 K. What is this in Fahrenheit?

11. At the beginning of the week, the temperature is measured at a chilly 40 °F. At the end of the week, the temperature has risen to 80 °F. Has the temperature doubled? Explain why of why not. (As a hint, consider look at question 10 and consider where the Fahrenheit scale begins).

Unit 1.2 Measuring Angles

Objective

To become familiar with angles, their measurements, conversion, and use in astronomy

Introduction

In geometry, any two connected points form a line. Two connected lines create an angle between them. Before the advent of telescopes (which allowed for detailed study of objects too small to see with the naked eye) astronomy as a science was limited to drawing star charts and carefully measuring the positions of planets against the background stars. Even with telescopes, careful studies of positions and measured angles allowed astronomers to determine the sizes of the planets and the distances between orbits.

Angles are measured in the unit of degrees. One full circle, rotation, or revolution contains 360 degrees. In much the same way as an hour is composed of smaller units called minutes, 1 degree is composed of smaller units called *arcminutes*. One arcminute is an incredibly small angle. Two lines which deviate by 1 arcminute would appear so close to one another that they be mistaken for one line to your naked eye. In fact, only after drawing the lines nearly 300 inches (25 feet) long would you see the ends of the lines separated by 1 inch, making an extremely long sliver of a triangle with a 1 inch long side.

Just as 1 minute of time is further broken down into 60 individual seconds, 1 arcminute is composed of 60 equally spaced subunits called *arcseconds*. Two lines which make an angle of 1 arcsecond to each other would have to be drawn out incredibly long before their deviation was noticeable. At 3.25 miles long, the ends of the two lines would lie one inch apart.

Angular diameter and angular separation are two critical concepts in astronomy and they both utilize the measurement of angles. The angular separation describes how distant two objects *appear* from one another in the sky. From our vantage point on Earth, the sky appears flat and two dimensional. The distance we see between stars is an angular separation, not an actual three dimensional, linear distance. For example, the three stars of Orion's belt appear very close to one another, making a straight line in the sky. In actuality, those three stars are each hundreds of light years away from the Earth and even further apart from one another. Their relative closeness (small angular separation) is an optical illusion. Along the same vein, when planets "align" during a conjunction and appear extremely close to one another in the sky, they are actually tens of millions of miles apart.

Angular diameters describe what percentage of your vision is occupied by an object. Human vision spans about 180° across. Something incredibly small – like the head of a pin from 100 yards away – takes up so little of your vision as to be invisible. The head of a pin viewed from a football field away would cover about 1 arcsecond of vision. The Great Wall of China – seen from orbit – is only about 20 arcminutes wide, still too small for the receptors in the eye to recognize.

Your fist – held out at arm's length – will cover roughly 10 degrees of your vision. In this way, angular diameter describes how large something appears to be rather than its absolute size.

Procedure

Measuring the angular separation between two lines requires a protractor. The standard protractor is a semi-circle marked off with ticks running from 0° to 180°. To measure the angle made by two lines, place the center of the protractor on the intersection point of the two lines.

Adjust the protractor until one of the lines is pointing directly at the angle of 0°. The second line will point to the reading representing the angle between the lines. It may make measurements easier to use a ruler and extend the length of the lines (which make the angle) so that you do not have to estimate where the measured line points.

Likewise, to draw an angle, begin by drawing two points and connecting them with a straight line. This will serve as the baseline. Place the center of the protractor at one end of the line, making sure that the line points to 0°. Mark a third point on your protractor at the desired angular separation in degrees and use the straight edge of a ruler to connect those points. Smaller units are one arcmin (1') and one arcsec (1").

Equations and Constants

$$1^\circ = 60'$$

$$1^\circ = 3600''$$

$$1' = \frac{1}{60}^\circ$$

$$1'' = \frac{1}{60}' = \frac{1}{3600}^\circ$$

NAME ______________________ ID ______________________

DUE DATE __________ LAB INSTRUCTOR ______________ SECTION __________

Worksheet # 1

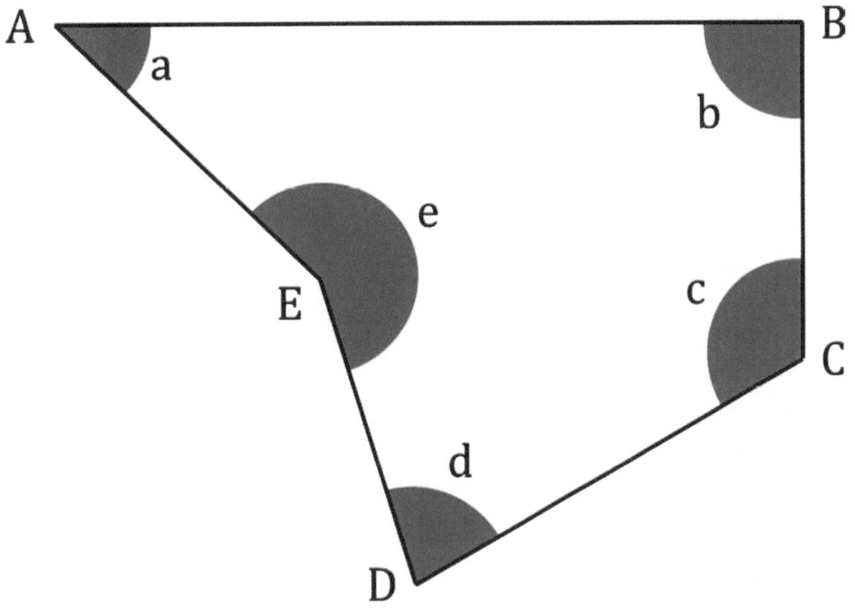

Below is a 5 sided geometric shape made up of lines AB, BC, CD, DE, and EA. At the point where the lines meet they make angles a, b, c, d, and e, with the angle highlighted in gray. Use a protractor to measure the indicated angles and give an answer in degrees.

1. Angle **a** (the angle between lines AB and AE)

2. Angle **b** (the angle between lines AB and BC)

3. Angle **c**

4. Angle **d**

5. Angle **e**

NAME ______________________ ID ______________________

DUE DATE __________ LAB INSTRUCTOR ______________ SECTION __________

Worksheet # 2

Using a protractor, draw a pair of lines intersecting at the listed angle.

1. 17°

2. 68°

3. 91°

4. 166°

5. 345°

NAME Emily ID 217066267

DUE DATE ________ LAB INSTRUCTOR A. Muzzin SECTION ________

Worksheet # 3

Use the conversions to determine the following angles.

1. In the Earth's sky, the moon has an angular diameter of 0.5°. What is the angular diameter of the Moon in arcminutes?

   0.5 x60 = 30 arcminutes

2. Through a telescope, two stars are separated by 2° 15' 35". What is their separation in arcseconds?

3. The stars Mizar and Alcor – both in the handle of the Big Dipper – are located 12' 21" from one another. What is that separation in arcseconds?

   21-12=9

   9x60 = 540

4. Arrange these angular diameters in order of largest to smallest angular diameter:

   a) 0.25°

   b) 90'

   c) 1000"

NAME ______________________ ID ______________________

DUE DATE __________ LAB INSTRUCTOR ______________ SECTION __________

Worksheet # 4

The four dashed circles below represent the orbits of the innermost planets, with S representing the position of the sun, V the position of Venus, E the position of Earth, and M the position of Mars at some given time.

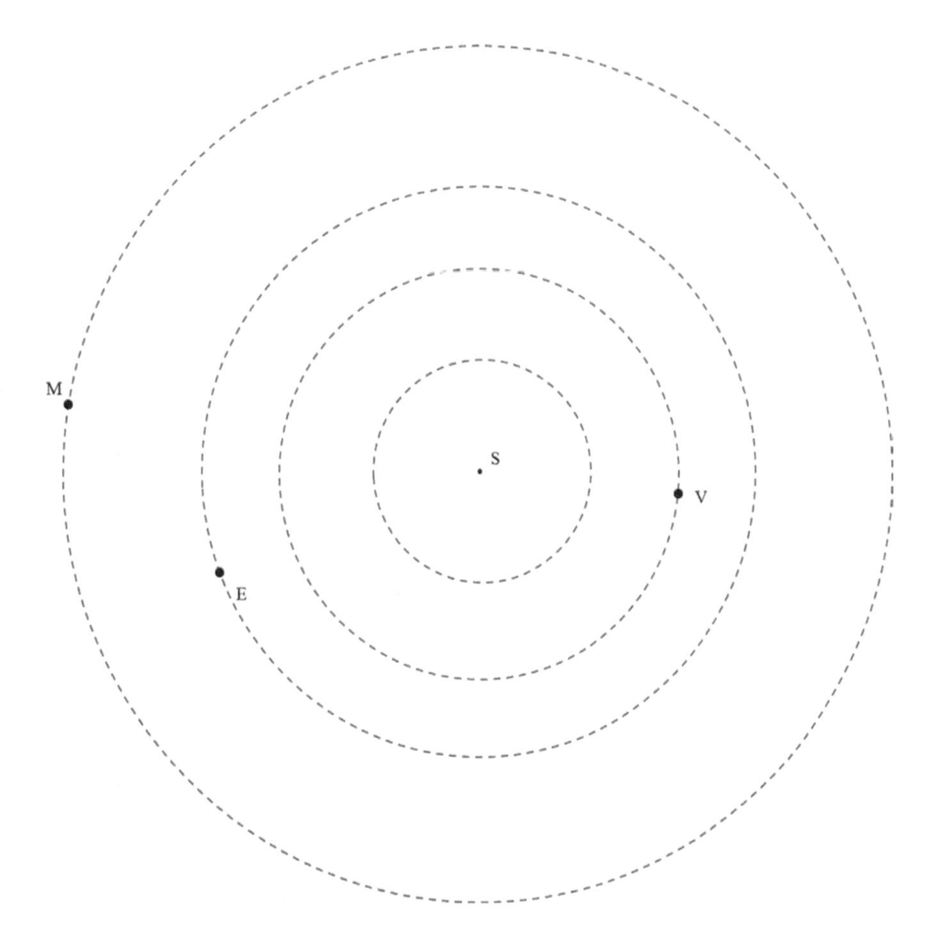

1. From the perspective of someone on Earth, how far apart would the Sun and Mars appear, in degrees? (What is the angle from M to E to S?) How far apart would the Sun and Venus appear, in degrees? (What is the angle from S to E to V?)

2. Someone on Earth determines that Mercury is 16° from the Sun. Mark Mercury's possible position in orbit with an ***M*** (Mercury's orbit is the innermost orbit.)

Motion on the Celestial Sphere

NAME Emily ID# ____________

DATE ____________

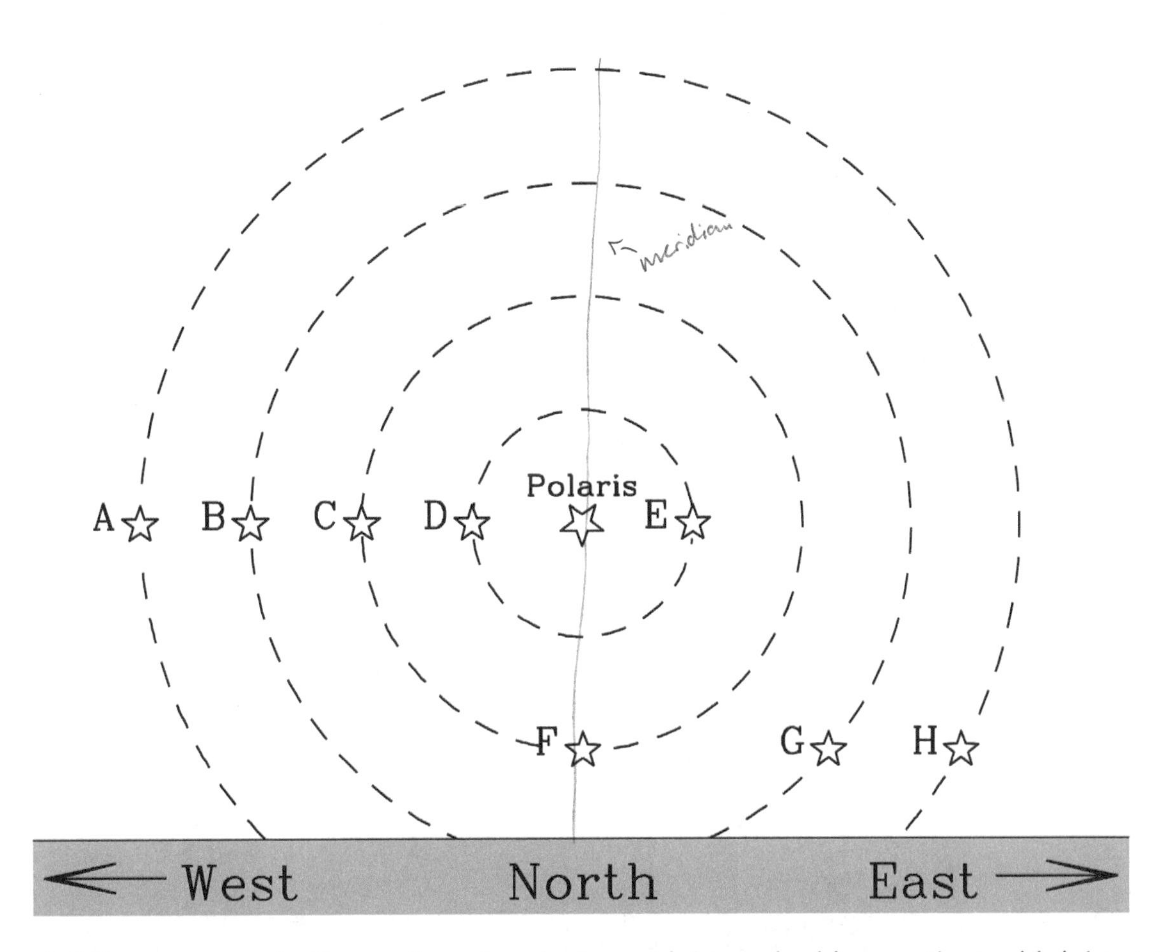

Figure 1: What someone in the northern hemisphere might see looking north at midnight at this time of year. The horizontal black line is your **horizon** (the grey area is part of the surface of the Earth). You can only see stars above your horizon.

Polaris (the North Star) is shown above, along with eight other stars. As the Earth rotates, all eight stars appear to move in circular paths around the North Celestial Pole (the NCP). The dashed lines show these circular paths.

1. Mark the location of the North Celestial Pole (NCP) in Figure 1. Polaris
2. Your **meridian** is the line on the celestial sphere going from due North through the NCP and continuing due South to stop at the point due South on your horizon. **Draw the part of your meridian visible** in Figure 1. Which star besides Polaris is on the meridian at the time shown? F
3. The Earth rotates so that the Sun rises above the Earth's horizon in the East and sets below the Earth's horizon in the West. Therefore, in which direction do these stars appear to move

on their circular paths around the NCP? *(Clockwise|Counterclockwise)* Draw an arrow at each star showing the direction that star appears to move at the time shown in Figure 1. [Note: These stars move through 360° in a 24-hour period.]

4. List any stars that are moving straight down towards the horizon: ______________
   List any stars that are moving straight up away from the horizon: ______________
   List any stars that are moving horizontally (i.e., parallel to the horizon): ______________
5. Will all these stars be above the horizon for the same length of time each day? *(Yes|No)*
6. Among the stars A, B, C, and D, which one will set first? ______A______
7. Stars that never set from a particular latitude are known as **circumpolar** stars. Which of the eight stars A through H are circumpolar? ______________
8. Figure 1 shows the sky at midnight. Six hours later (6 A.M.) the Sun will rise. Which of the stars A through H will be above the horizon at sunrise? ______________ At 6 P.M., six hours earlier than the time shown in Figure 1, the Sun had set and the sky had just gotten dark enough to see stars. Which of the stars A through H was visible then? ______________
9. Mark the meridian and the zenith for the tiny stick figure on the Earth in Figure 2.

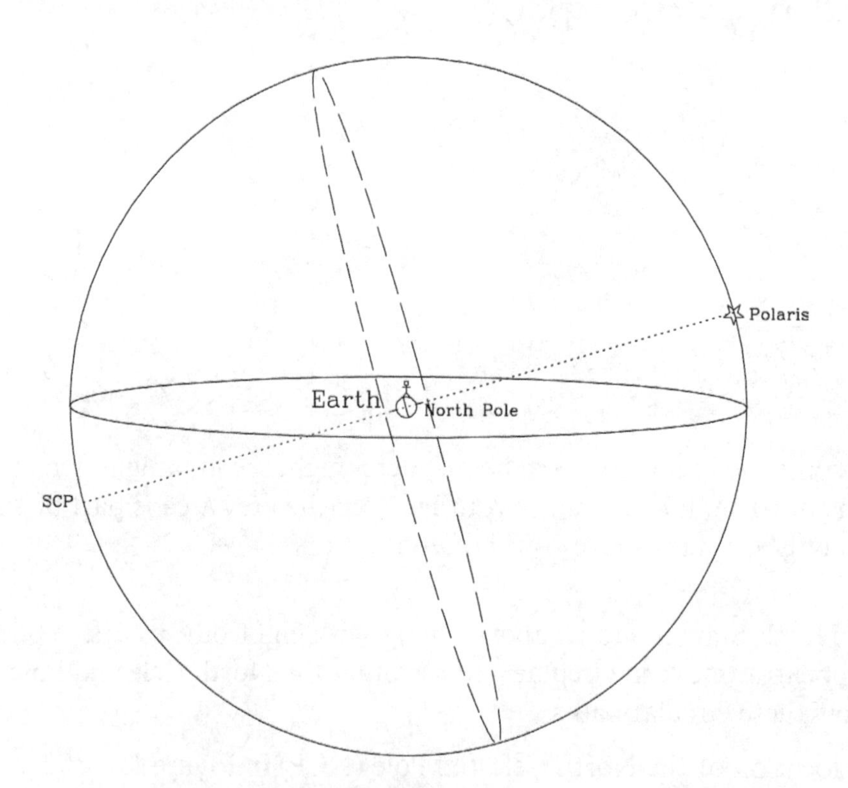

Figure 2: The same situation as shown in Figure 1, drawn in terms of our model of the **celestial sphere:** an imaginary spherical surface surrounding the rotating Earth. In reality, stars are located at many different distances from Earth, but we can imagine them as being projected onto the celestial sphere. Also, because the Earth spins on its axis (the line connecting Earth's North Pole and South Pole), the stars appear to move. However, relative to our point of view on Earth, the Earth doesn't move and we can instead imagine the celestial sphere rotating around the Earth on the same axis (the dotted line).

10. Mark the horizon for the stick figure in Figure 2. Indicate which parts of Figure 2 are above the horizon for that stick figure by shading only the parts that are below the horizon. (Hint: We are viewing the stick figure from slightly above its horizon.)
11. Draw on Figure 2 the circular path on the sky of any one of the stars in Figure 1, including arrows to indicate the direction of motion of the star on its path.
12. What is the dashed line girding the celestial sphere in Figure 2? _______________
13. (**Optional**) Shade in the region of the celestial sphere where stars that are *never visible* from the stick figure's location are found.

Motions of the Moon

NAME ______________________ DATE ______________________

Definitions

Define the following terms in your own words:

1. Eclipse

2. Conservation of Angular Momentum

3. Lunar Month (Synodic Month)

4. Sidereal Month

5. Roche Limit

6. Tidal Locking

NAME ______________________________ DATE ______________________________

Important Facts

1. Sketch the alignment of the Sun, Moon, and Earth for the following:
   a. a solar eclipse

   b. a lunar eclipse

   c. spring tides

   d. neap tides

2. a. The Moon's orbital period (sidereal month) is _______________days
   b. The Moon's phase period (lunar month) is _______________days
3. a. At what phase of the Moon does a solar eclipse occur? _______________
   b. At what phase of the Moon does a lunar eclipse occur? _______________
4. What is the current rate at which the Moon recedes from Earth under tidal influences? _______________cm/yr
5. What are the only two worlds in the Solar System that are known to be tidally-locked to each other? _______________and_______________

NAME ______________________ DATE ______________________

6. Complete the following table of lunar phases:

Phase	Sketch	Moonrise/Moonset	Orientation to Sun
New		**Moonrise:** **Moonset:**	
Waxing Crescent		**Moonrise:** **Moonset:**	
First Quarter		**Moonrise:** **Moonset:**	
Waxing Gibbous		**Moonrise:** Afternoon **Moonset:** After midnight	Between 90° to 180° east of the Sun
Full		**Moonrise:** **Moonset:**	
Waning Gibbous		**Moonrise:** **Moonset:**	
Last Quarter		**Moonrise:** **Moonset:**	
Waning Crescent		**Moonrise:** After midnight **Moonset:** Afternoon	< 90° west of the Sun

NAME ______________________________ DATE ______________________________

Critical Thinking Questions

1. Why is the lunar month different from the sidereal month? Draw a sketch to accompany your explanation.

2. Why are some solar eclipses total while others are annular? Were total eclipses or annular eclipses more common in the distant past? Which will be more common in the distant future? Explain your answers.

3. If the Moon is receding from Earth, then the angular momentum in its orbit is increasing. Where does it get this additional angular momentum?

Lunacy: Moon Phase Exercise

NAME ______________________________ DATE ______________________________

1. Complete the following moon phase chart with rising, setting and culmination times (sunset, sunrise, noon and midnight). Refer to the moon phase diagram on the preceding page.

	New Moon	First Quar.	Full Moon	Last Quar.
Rising				
Culmination				
Setting				

2. What phase occurs at position 4?

   What phase occurs at position 1?

   What phase occurs at position 2?

   What phase occurs at position 3?

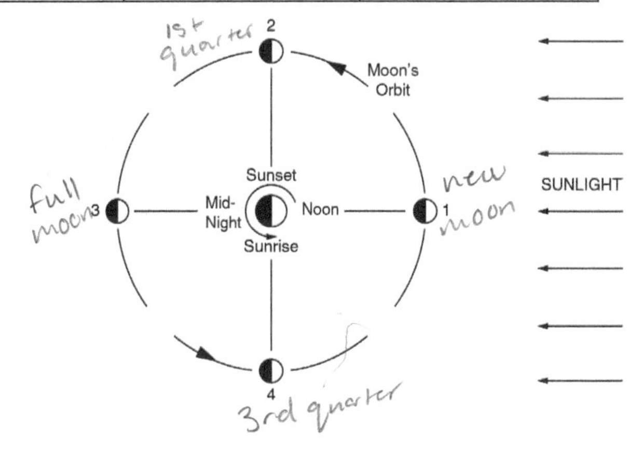

3. What moon phase can create a **lunar eclipse**? Full Sketch the moon, earth and sun in alignment and include the earth shadow and moon shadow. (Use the back of the page.)
4. What moon phase can create a **solar eclipse**? new Sketch the moon, earth and sun in alignment and include the earth shadow and moon shadow. (Use the back of the page.)
5. Why doesn't a lunar and solar eclipse occur every month?
6. What time does the Waxing Gibbous Moon **RISE**? (Use moon phase chart.) ____ Can you see the Waxing Gibbous Moon in **DAYLIGHT?**
7. You are planning a "star party" and the sky must be dark and "moonless" (when the moon is below the horizon between **sunset to midnight**). What moon phases will be ideal for this star party? ______________
8. At midnight where would the Last Quarter Moon be? __________
9. An occultation of Jupiter by the Waning Gibbous Moon will occur at 7:15 pm, local time. Can you see the event? (Where is the Waning Gibbous Moon at 7:15 pm?) ____

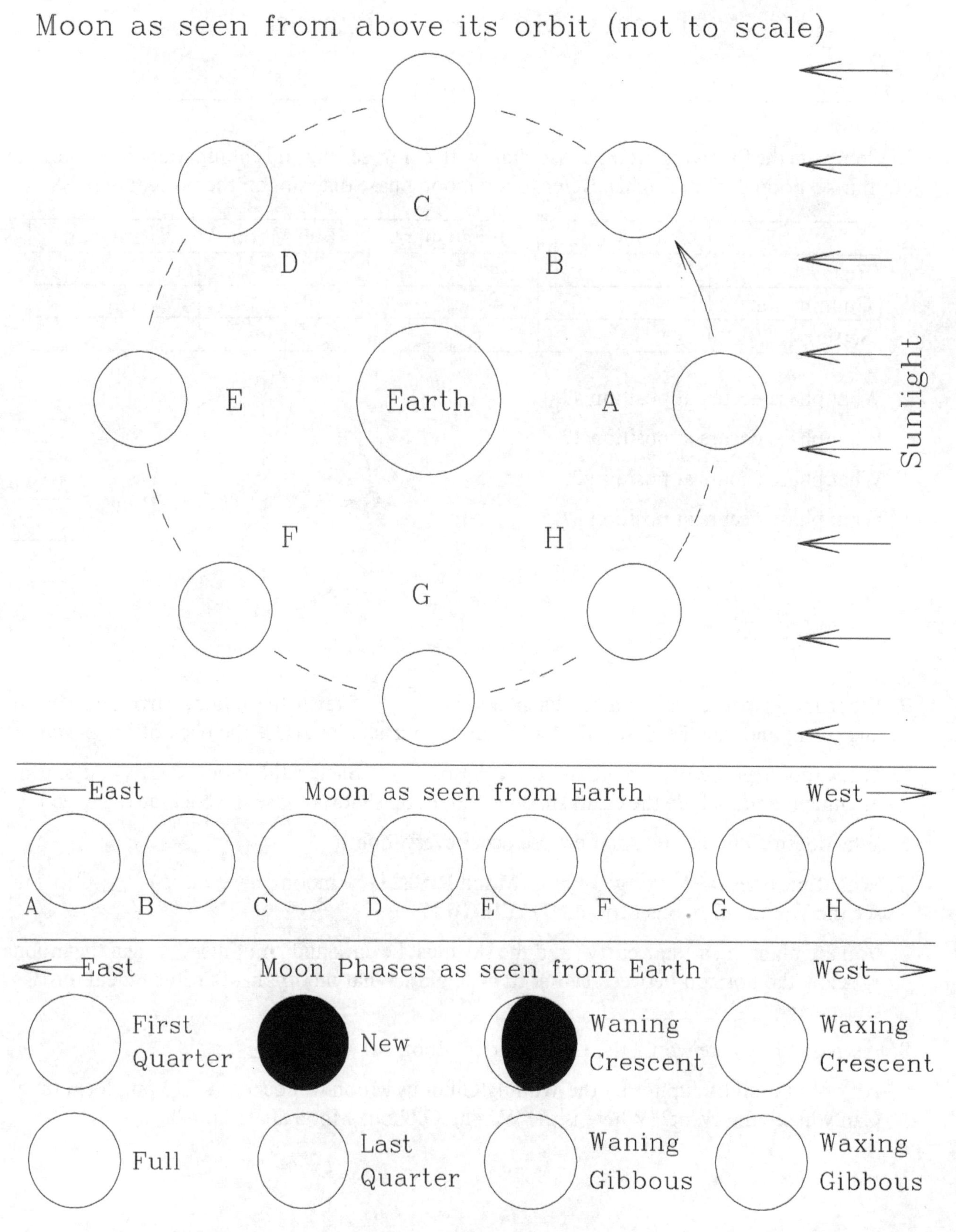

Figure 1: (Top) The Moon as seen from above its orbit. (Middle) The Moon as seen from Earth. (Bottom) Phases of the Moon. The Full, New, and Waning Crescent phases have been drawn in for you.

The Phases of Venus

TEAM #________________________

Please print your name and sign next to it (only those present).

Leader: (C)____________________ ____________________

Explorer: (D)____________________ ____________________

Skeptic: (A)____________________ ____________________

Recorder: (B)____________________ ____________________

Learning Objectives

1. Use the Ptolemaic model to predict the phases of the Venus.
2. Use the Copernican model to predict the phases of the Venus.
3. Understand how Galileo's observation of the phases of Venus helps to select between these two models.

Introduction: As the first scientist to undertake detailed astronomical study with the aid of a telescope, Galileo Galilei (1564-1642) was able to make a number of important discoveries that greatly influenced our understanding of the universe. These observations were published in 1610 in his book *The Starry Messenger*, which was written in a style that made it accessible to a wide audience. One of his observations was that Venus exhibits phases much like our Moon.

Part I: Venus in the Ptolemaic Model

In the Ptolemaic model, Venus moves on an epicycle whose center is always directly between Earth and the Sun. The figure on page 35 shows Venus in a series of locations around its epicycle. Your task is to sketch the appearance of Venus (as observed through a telescope from Earth), when Venus is at each of the five indicated positions. Use the following procedure:

- For each of the five positions along the epicycle (i.e., the *top half* of the page) shade in the dark half of Venus.
- For each of these positions sketch (in the spaces at the *bottom* of the page) what Venus would look like as seen from Earth. In your sketches, shade the portion that would appear dark and the leave alone the part that would appear bright.

1. Is there any location along the epicycle at which Venus would appear as a near fully lit disk? (If there is, indicate which one.)

Part II: Venus in the Copernican Model

In the Copernican model, Venus moves around the Sun in an orbit that is only 72% the size of Earth's orbit. The figure on page 36 shows Venus at five different locations around the Sun. Your task is to repeat the procedure from Part I to sketch the appearance of Venus as seen from Earth when Venus is observed at each of these locations.

2. Is there any location along its orbit at which Venus would appear as a near fully lit disk? (If there is, indicate which one.)

3. Galileo observed that Venus goes through a complete set of phases including full phase. With which system of planetary motion, Ptolemaic or Copernican, is this observation consistent? Explain.

4. Do these observations of the phases of Venus necessarily confirm the heliocentric model? (Hint: think about Tycho's model in which the Sun orbits the Earth but Venus orbits the Sun.) Explain your answer.

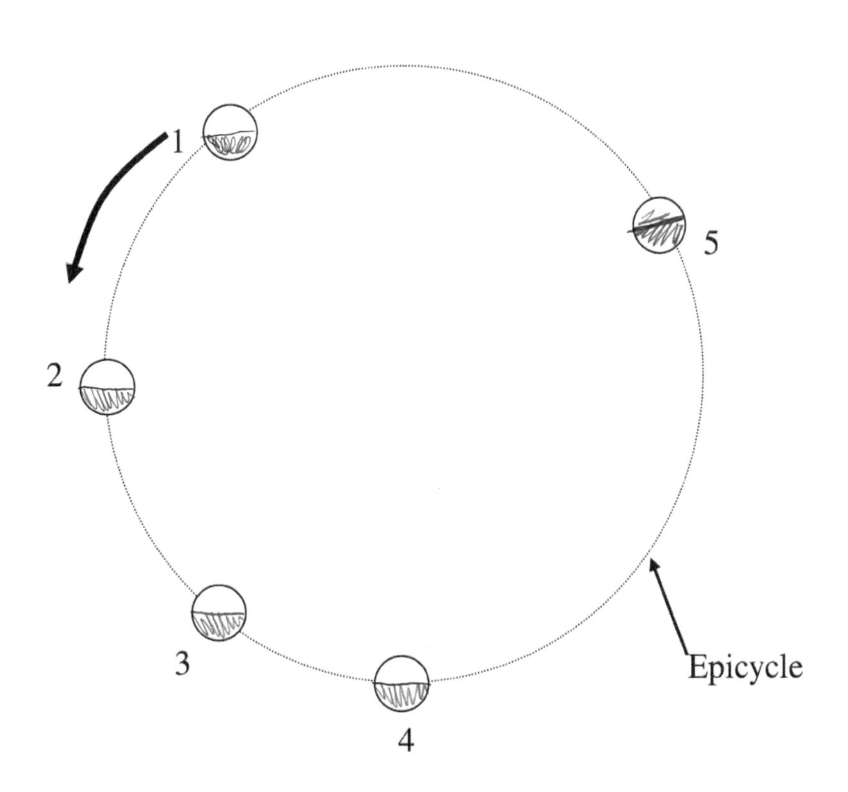

Sketch Venus's appearance as seen from Earth at the five locations shown above.

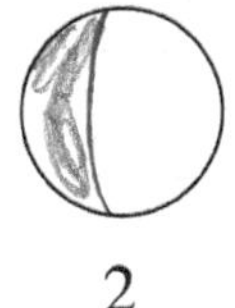

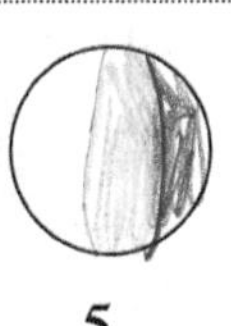

1 2 3 4 5

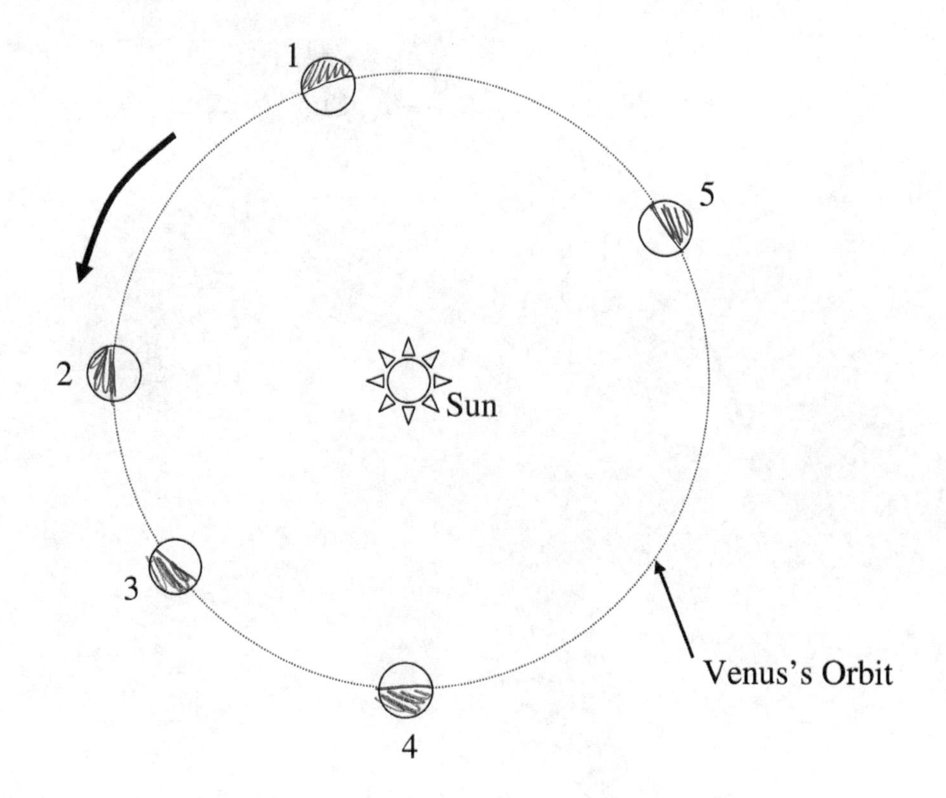

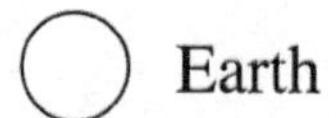

Sketch Venus's appearance as seen from Earth at the five locations shown above.

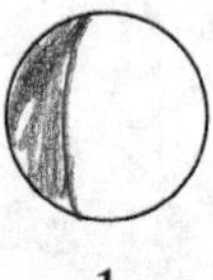
1

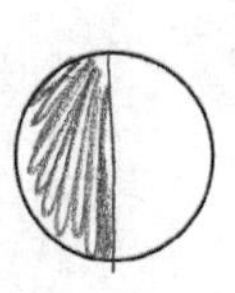
2

3

4

5

Mapping the Solar System from Earth

TEAM #____________________________

Please print your name and sign next to it (only those present).

Leader: (D)____________________________ ____________________________

Explorer: (A)____________________________ ____________________________

Skeptic: (B)____________________________ ____________________________

Recorder: (C)____________________________ ____________________________

Learning Objectives

1. Comprehend that the observer's position on Earth makes particular objects in the sky visible at specific times.
2. Analyze the rotation of an Earth observer to predict the rising & setting times of sky objects.
3. Synthesize heliocentric object locations and interpret to a geocentric perspective.
4. Synthesize geocentric object positions and interpret to a heliocentric perspective.

Background: Some newspapers and science magazines, such as *Sky and Telescope*, provide sky charts that describe what sky objects are visible at different times. These typically include

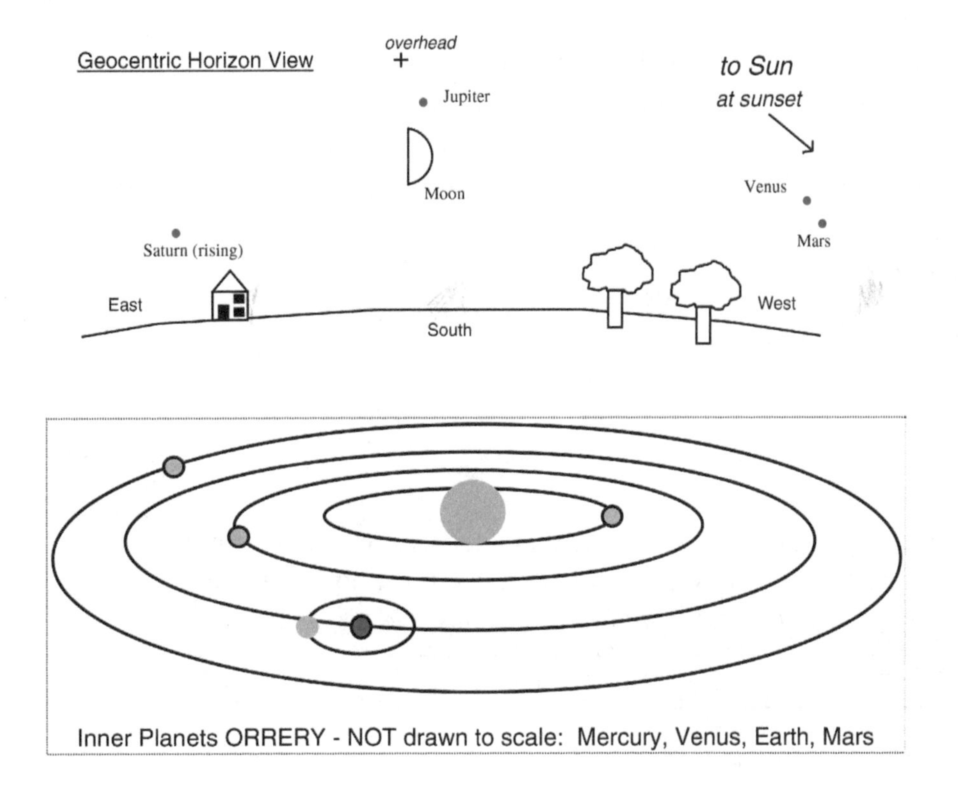

prominent stars, bright planets, and the Moon. There are two principle maps provided to readers: (1) a geocentric horizon view and (2) a heliocentric orrery view. The *geocentric* perspective is the view from Earth looking up into the southern sky. The *heliocentric* perspective is the view of the Solar System looking down from above. From above, the plants orbit and spin counter-clockwise (except Venus, which appears to spin backwards).

Part I: Rising and Setting Times

As seen from above, Earth appears to rotate counterclockwise. Figure I-a shows a top view of Earth and an observer at noon. Note that our Sun appears overhead when standing at the equator.

Figure I-a: Observer Positions on Earth [Observer is at Equator]

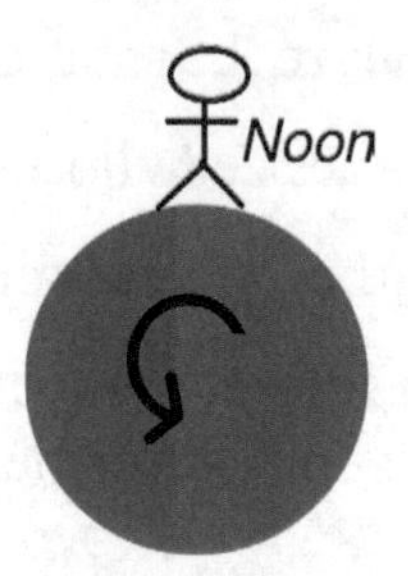

1. In Figure I-a, sketch and label the positions of the observer at midnight, 6 pm (sunset) and 6 am (sunrise).

2. Consider Figure I-b, which shows Earth, Moon, Mars, and Venus. At what time would each of these sky objects be overhead? Remember that Earth spins counter-clockwise when viewed from above. [*Hint: Make use of Figure I-a*]

   Time Overhead:

   Venus: ____________________

   Moon: ____________________

   Mars: ____________________

Figure I-b: Orrery

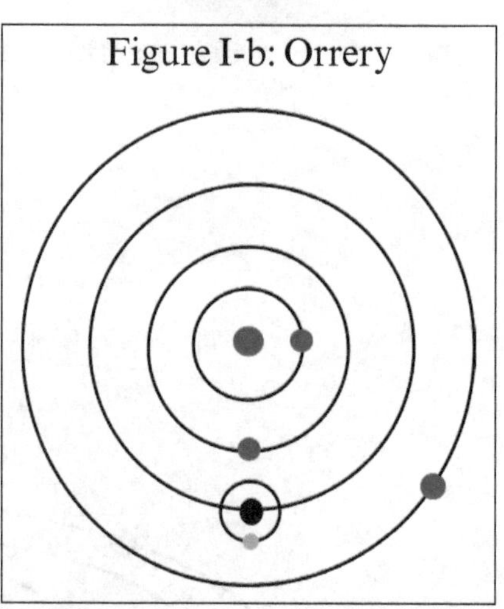

3. If Earth spins in 24 hours, it means that each sky object is visible for about 12 hours. What time will the sky objects shown in Figure I-b rise and set? Complete the table below? *Each member of your team should fill in the data for one sky object.*

Sky Object	Rise Time	Time Overhead	Set Time
Sun			
Venus			
Moon			
Mars			

4. Using complete sentences, explain why our Sun is not visible at midnight. Add a sketch of Earth, Sun, and observer in the space provided to support your explanation.

Narrative	Sketch

Part II: Converting Geocentric to Heliocentric

5. Figure II-a shows the horizon view of the first quarter Moon and Saturn visible at sunset. On the orrery shown in Figure II-b, sketch and label the position of Jupiter, Moon and Saturn. Use an arrow to indicate the direction to our Sun. Start by indicating the position of the observer at sunset. After completing the diagram, complete the table.

Sky Object	Rise Time	Set Time
Sun		
Jupiter		
Moon		
Saturn		

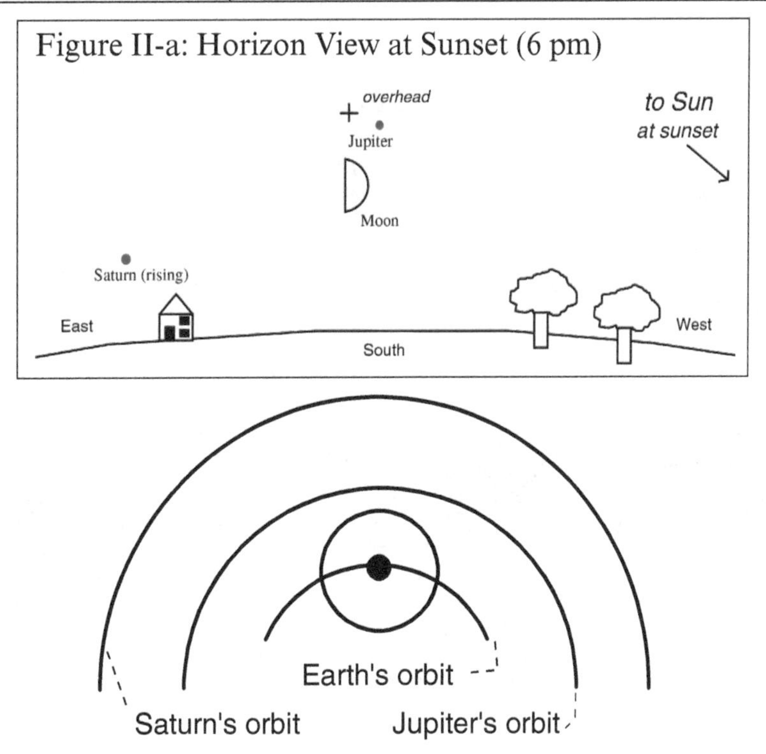

Orrery Not Drawn to Scale !!

Figure II-b

6. If Neptune is visible overhead in the southern sky at sunrise (6 am) sketch the relative positions of Sun, Earth, Neptune, and observer in an orrery in the space below.

Part III: Converting Heliocentric to Geocentric

7. Figure III-a shows the position of Mercury, Venus, Earth, Mars, and Moon. On the horizon diagram, Figure III-b, sketch and label the positions of Mercury, Venus, Mars, a comet, and Moon at <u>midnight</u>.

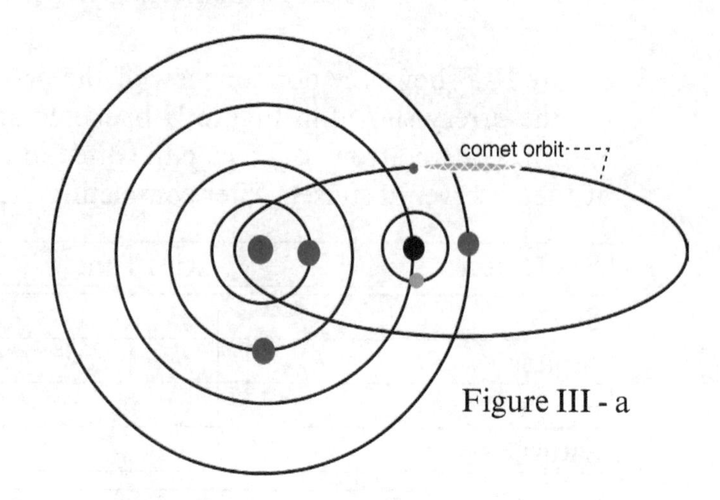

Figure III - a

<u>Figure III-b: Geocentric Horizon View at Midnight</u>

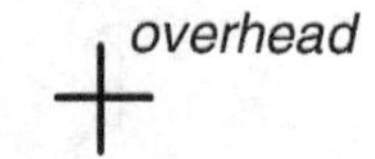

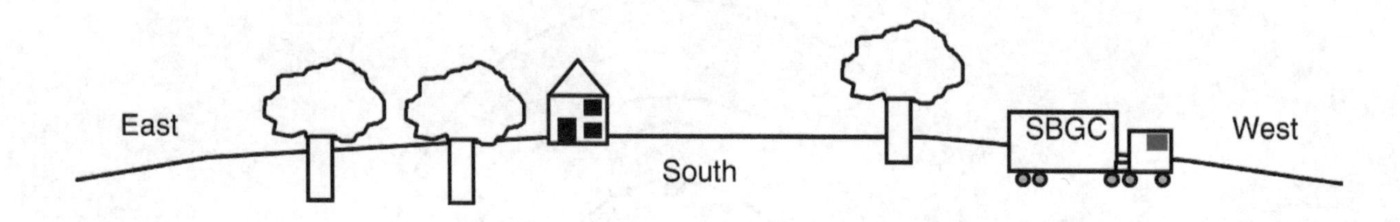

8. Venus is often called the *morning star* or the *evening star*. Why is it never seen at midnight?

Part IV: Current Events

9. Your instructor will provide you with a magazine photocopy or software print-out of the current sky or orrery. Convert the given current sky to an orrery OR convert the current orrery to a horizon view (you can select the time). Make your sketches in the space below. Be certain to label every item carefully.

10. Describe and sketch the night sky if you were to go outside at midnight tonight (*assume there are no clouds*).

Moon Phase Poster

	New Moon	Day 0 and Day 28
	Waxing Crescent	Days 1-6
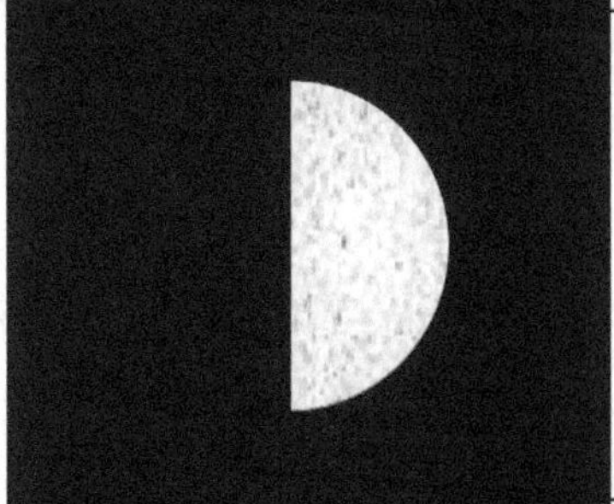	First Quarter	Day 7
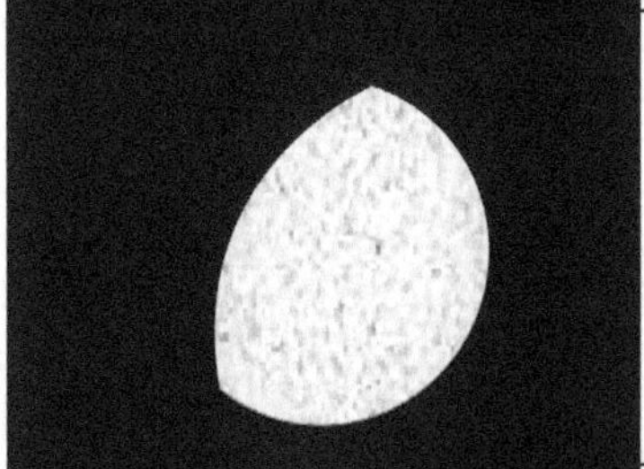	Waxing Gibbous	Days 8-13
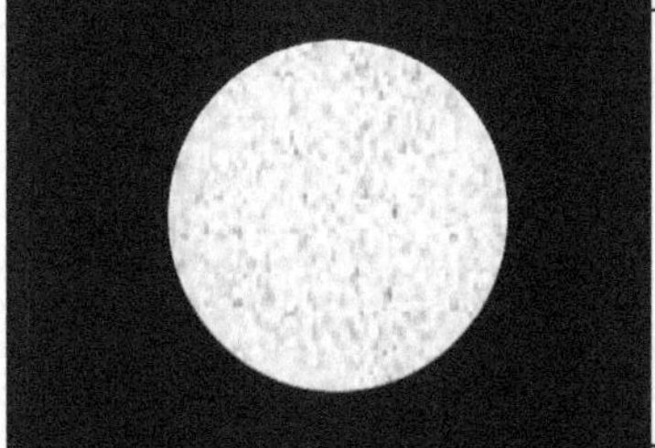	Full Moon	Day 14
	Waning Gibbous	Days 15-20

	Third Quarter	Day 21
	Waning Crescent	Days 22-27

The Seasons and the Sun's Path on the Celestial Sphere

NAME ______________________ ID# ______________________

DATE ______________________

Figure 1 on the next page shows the visible hemisphere of the celestial sphere for someone located at latitude 45 degrees north of the equator. If you are the stick figure, then the grey area is the part of the Earth's surface visible to you; the grey area ends at **your horizon.** The **cardinal directions** N(orth), E(ast), S(outh), and W(est) are indicated on your horizon.

The Sun's path through the sky on the **equinoxes** is drawn as the long-dashed line, passing through point b. The Sun's paths on the **solstices** are drawn as the two short-dashed lines, one passing through point a and the other through point c. Ignore the dotted line for now.

1. Mark your horizon and your **zenith** on Figure 1.
2. The Sun is shown crossing the meridian on a specific day of the year. What day is that? december solstice (longest day of the year) Label that position of the Sun on Figure 1 with that date.
3. Draw the Sun crossing the meridian on the remaining solstices and equinoxes, and label each position of the Sun with the date it would be seen there.
4. On what days does the Sun rise due East and set due West? equinoxes?
5. Is the Sun ever directly overhead at this person's location? If so, on what day(s)?

   ____

   The following two questions refer to the Northern Hemisphere only.
6. Between what days does the Sun rise north of East and set north of West? What season(s) occur at that time of year? summer
7. In what season(s) does the Sun rise south of East and set south of West? equanoxes + winter

Figure 2 shows the southern half of the sky visible from the position of the stick figure in Figure 1. The **visible** celestial hemisphere has been cut in half on the dotted line (which is the same in both figures) and its southern half is shown. The Earth (grey) ends at the horizon, and the zenith is marked. The long-dashed and short-dashed lines are the same as in Figure 1.

8. Draw in the part of your meridian visible in Figure 2.
9. Add the labels a, b, and c to the appropriate locations in Figure 2.
10. Shade in the parts of the sky in Figure 2 in which the Sun will never be seen.
11. **(Optional)** If you observe the Sun move along the dotted line in Figure 1 over the course of a day, where on Earth are you located, and what day of the year is it?

    ____
12. **(Optional)** What is the maximum altitude reached by the Sun on an equinox, and how does this relate to the latitude of the stick figure? (Hint: The zenith is 90° from the horizon.)

    ____

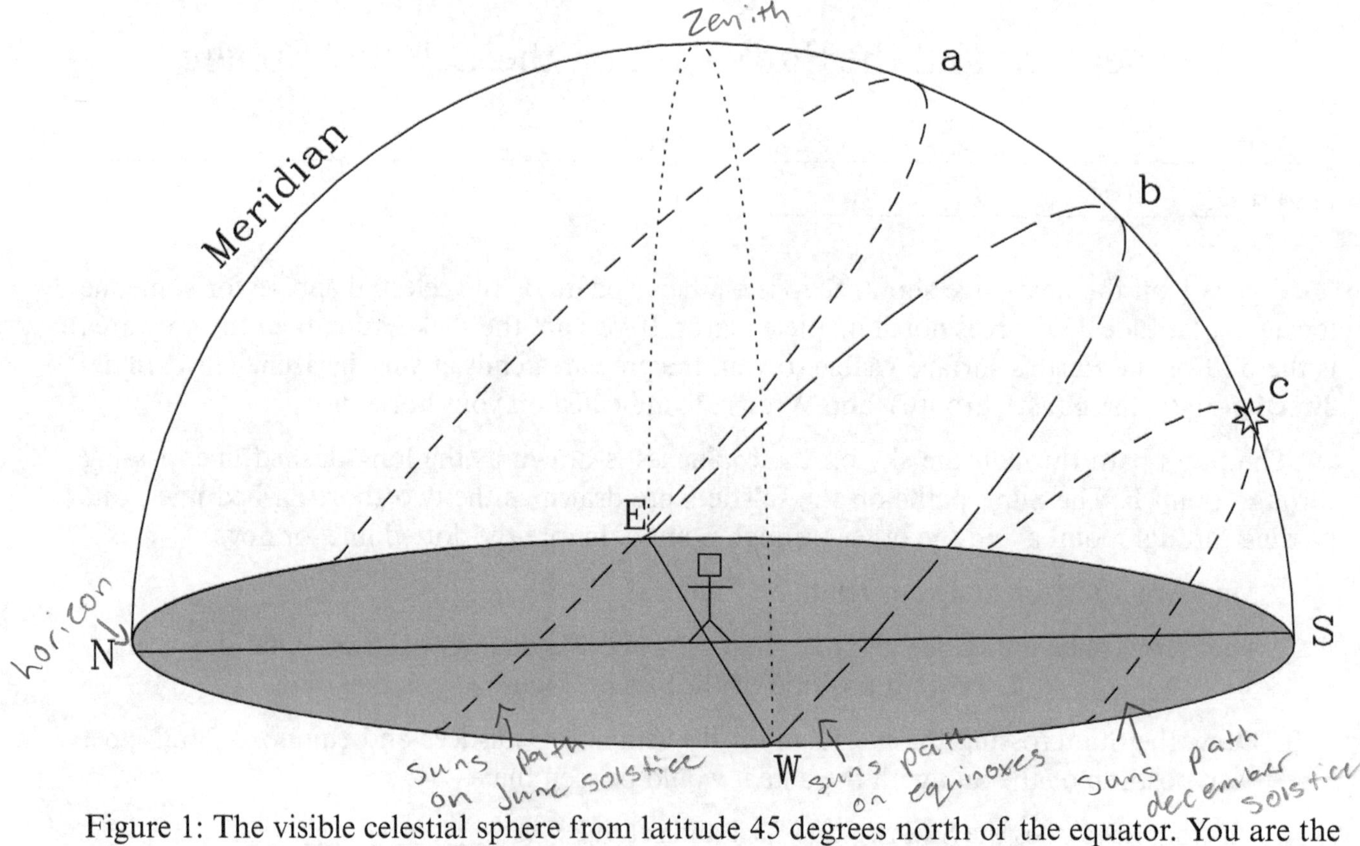

Figure 1: The visible celestial sphere from latitude 45 degrees north of the equator. You are the stick figure, and the part of the Earth you can see (grey) ends at your horizon. The Sun (star) follows the long-dashed line on the equinoxes and the short-dashed lines on the solstices.

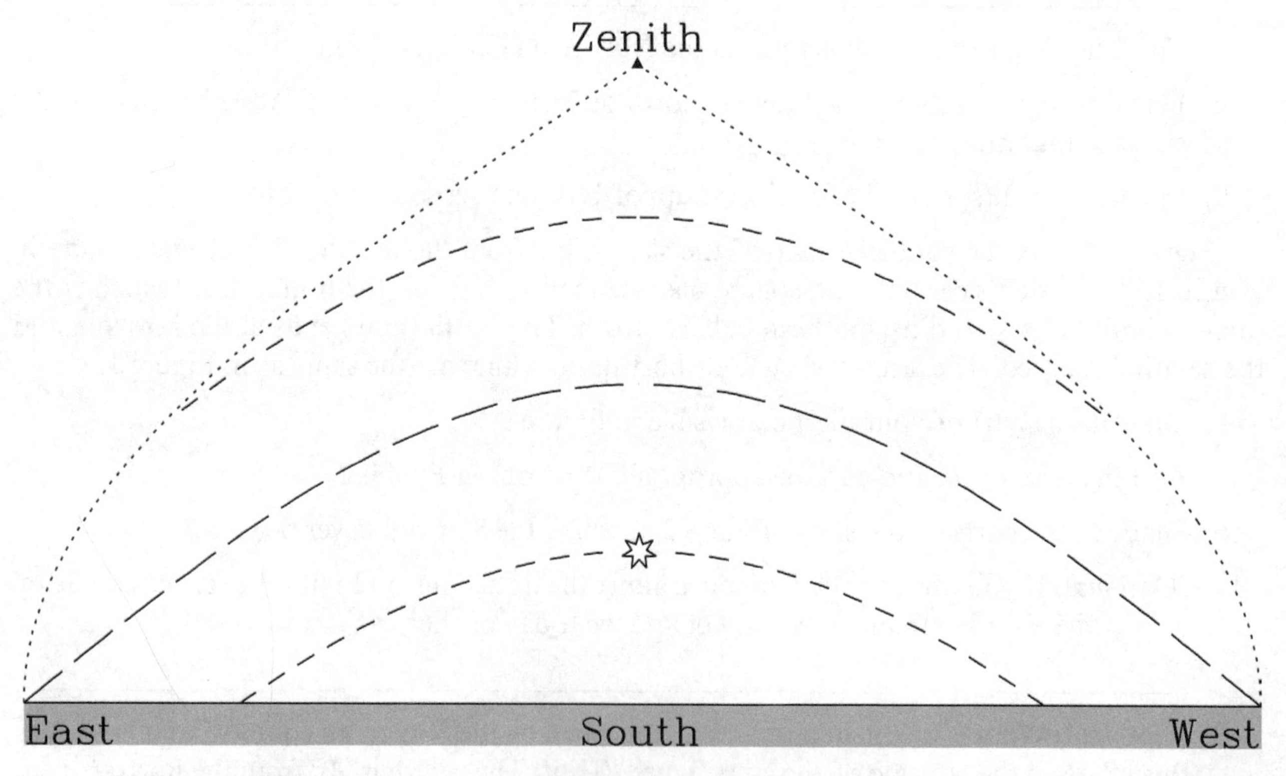

Figure 2: The southern half of the sky visible from the position of the stick figure in Figure 1. The visible celestial sphere in Figure 1 has been cut on the dotted line and the southern half of the sky is shown here. The lines and Sun symbol are the same as in Figure 1.

Force, Acceleration, and Gravity

NAME Emily **ID#** 217066267 **DATE** October 4, 2019

Newton's Second Law of Motion

Isaac Newton published his **Three Laws of Motion** in 1687. To a very good approximation, these laws describe the motion of all objects in the universe.

Newton's Second Law of Motion can be stated as:

$$\text{Acceleration of object } (a) = \frac{\text{Force applied to object } (F)}{\text{mass of object } (m)}$$

Force is how hard something pushes or pulls on another thing. Acceleration is **how fast velocity (v) changes with time**. For example, if you are in a car that goes from 0 to 100 kph in 10 seconds, you experienced an acceleration of 10 kph per second.

1. Suppose you kick a football as hard as you can and then kick a sandbag of the same size with the same force.

   How far will the football go? far → different masses

   How far will the sandbag go relative to the football? not as far

2. Let's compare your common sense to Newton's Second Law. If the sandbag has 50 times the **mass** (m) of the football, which object will have a larger acceleration? football

3. Your foot touches the football and the sandbag for the same length of time (t). The velocity you give each object is, therefore:

   $$v = at.$$

   How much greater velocity will the football have than the sandbag? __________

Newton's Universal Law of Gravitation

Newton worked out that the force of gravity F_g between two objects with masses m_1 and m_2 depends on the product of those masses divided by the square of the distance d between them:

$$F_g = \frac{Gm_1m_2}{d^2}$$

where G is a **constant** whose numerical value depends on the units you use. (If you measure the force between two specific objects in **Earth-pounds**, 1 Earth-pound being the force needed to hold up a weight of 1 pound in Earth's gravity, you must get the same answer for F_g in Earth-pounds no matter what units you use for mass and distance.) We will work in units where $G = 1$.

Figure 1 shows four pairs of asteroids with masses and distances given in units where $G = 1$. The asteroids have been held in place until this instant, so they have zero velocity. Each asteroid feels a gravitational force (and thus an acceleration) only from the other member of the pair.

4. **Estimate** or **guess** which pair of asteroids in Figure 1 experiences the strongest gravitational force? (Don't do any calculations!) ____________

5. **Calculate** the gravitational forces between the pairs of asteroids in Figure 1. For example, if there was a pair of asteroids labeled Z with $m_1 = 3$, $m_2 = 2$, and $d = 1$, the force between those two asteroids would be $F_Z = (m_1 \times m_2)/d^2 = (3 \times 2)/1^2 = 6$ units.

   $F_A =$ $\qquad$ $F_B =$ $\qquad$ $F_C =$ $\qquad$ $F_D =$

6. Use those results to complete this sentence: The weakest gravitational force is felt by __B__, the next weakest by __A__, the second strongest by __C__, and the strongest by __D__.

Acceleration Due to Gravity

We've worked out the gravitational force acting between each pair of asteroids, but what we observe is the acceleration of each asteroid. We can combine Newton's Second Law of Motion and Newton's Law of Universal Gravitation to find the acceleration of each asteroid in a pair:

$$a_1 = \frac{F_g}{m_1} = \frac{Gm_1m_2}{m_1d^2} = \frac{Gm_2}{d^2} \quad \text{and} \quad a_2 = \frac{F_g}{m_2} = \frac{Gm_1m_2}{m_2d^2} = \frac{Gm_1}{d^2}$$

Note that the acceleration of asteroid 1 does not depend on the mass of asteroid 1 and that the acceleration of asteroid 2 does not depend on the mass of asteroid 2. **The acceleration of an object due to the gravity of a second object depends only on the mass of the second object and the distance between the objects.**

7. Which asteroid in Figure 1 do you think experiences the largest acceleration? ____________

8. Calculate the acceleration of every asteroid in Figure 1. Express your answer as fractions.

   A: $a_1 =$ $\qquad$ B: $a_1 =$ $\qquad$ C: $a_1 =$ $\qquad$ D: $a_1 =$

   A: $a_2 =$ $\qquad$ B: $a_2 =$ $\qquad$ C: $a_2 =$ $\qquad$ D: $a_2 =$

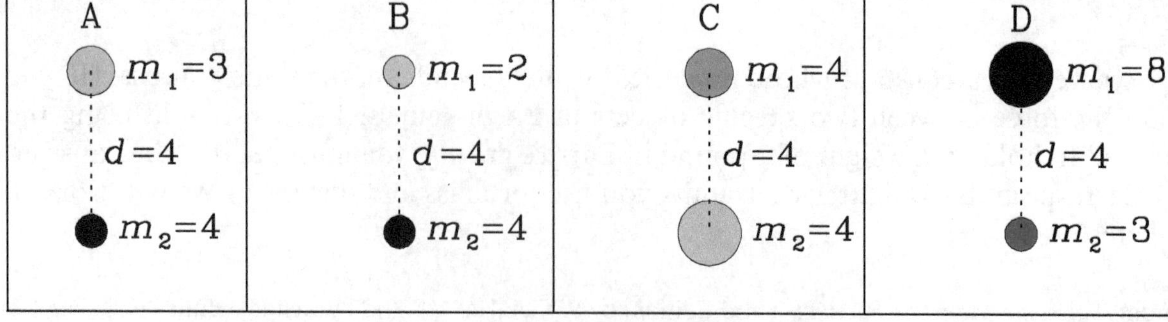

Figure 1: Four pairs of asteroids, with their masses and the distances between them given.

NAME ________________ ID# ______________ DATE ____________________

9. Which asteroid in Figure 1 has the smallest acceleration? ______________

   Which asteroid in Figure 1 has the largest acceleration? ______________

10. Was your answer to Question 7 correct? ______________

    Whether you were correct or not, explain how you would choose the correct answer next time:

    ___

    ___

11. Look at asteroids A1, B1, C1, and C2, and the asteroids each of them was paired with. What two things do asteroids A1, B1, C1, and C2 have in common? (Hint: The two things are related.)

    ___

    ___

12. If I replace asteroid A1, B1, C1, or C2 with Asteroid M, which has mass *M*, what will the acceleration of Asteroid M be, and how do you know?

    ___

    ___

13. Figure 2 shows four small asteroids next to one very large asteroid. Which small asteroid (A, B, C, or D) has the largest acceleration at the instant shown?

    A ________ B ________ C ________ D ________ All will have the same acceleration. ________

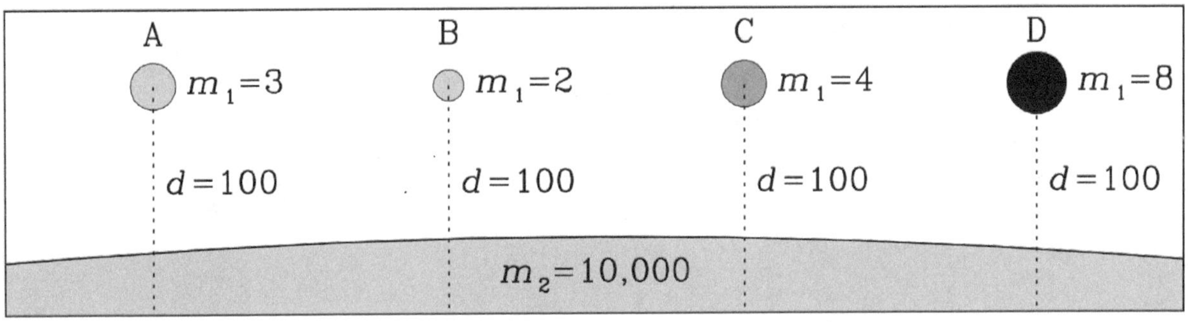

A $m_1=3$ $d=100$

B $m_1=2$ $d=100$

C $m_1=4$ $d=100$

D $m_1=8$ $d=100$

$m_2=10{,}000$

Figure 2: Four small objects next to one very large object.

14. Figure 2 could represent four spheres held above the surface of the Earth. The asteroid(s) or sphere(s) with the largest acceleration will be the first to collide with the object of mass m_2 (call it M2). (The objects are all the same distance from the surface and center of M2, and the acceleration of M2 is so small it can be ignored.) Based on your answer to Question 13, and on your experience here on Earth, which object(s) will collide with M2 first?

The following two questions summarize the difference between the **force of gravity** and the **acceleration due to gravity**.

15. The Earth pulls on you with a gravitational force which is... (choose one)
    a. larger than the gravitational force with which you pull on the Earth.
    b. equal to the gravitational force with which you pull on the Earth.
    c. smaller than the gravitational force with which you pull on the Earth.

16. Because of the force of gravity between you and the Earth, if you step off a table you will be accelerated toward the Earth. At the same time... (choose one)
    a. the Earth will be accelerated toward you much more than you are accelerated toward it.
    b. the Earth's acceleration toward you will exactly equal your acceleration toward it.
    c. the Earth will be accelerated toward you much less than you are accelerated toward it.

17. (**Optional**) Each asteroid pair in Figure 1 was held in place with rockets until this instant. Gravity will pull each pair of asteroids together. Which asteroid pair will collide first, and why?

    ____

    ____

Kepler's Laws and Elliptical Orbits

NAME ______________________________ ID# ______________________________

DATE ______________________________

Introduction

Tycho Brahe observed the regularly changing positions of the planets relative to the stars over several decades. **Johannes Kepler** used these observations to determine how objects orbit the Sun and summarized his results in three Laws. Newton and Einstein later showed that these Laws are only approximations, but they are accurate enough for us to use in many cases.

Kepler's First Law

An object which orbits the Sun moves along an ellipse with the Sun at one focus.

An ellipse can be thought of as a stretched circle. An ellipse has two **foci** (the plural of focus) which are located inside the ellipse on its **major axis** (the longest direction of the ellipse). The **center of the ellipse** is exactly halfway between the foci.

An ellipse's **eccentricity** (e) is the distance between its foci divided by the length of its major axis. An ellipse with $e = 0$ is just a circle. An ellipse with $e = 1$ is infinitely long and thin.

The **semi-major axis** (a) is half the length of the major axis. For a circle, a equals the radius (r).

Use the above information to answer the following questions for the elliptical orbit in Figure 1.

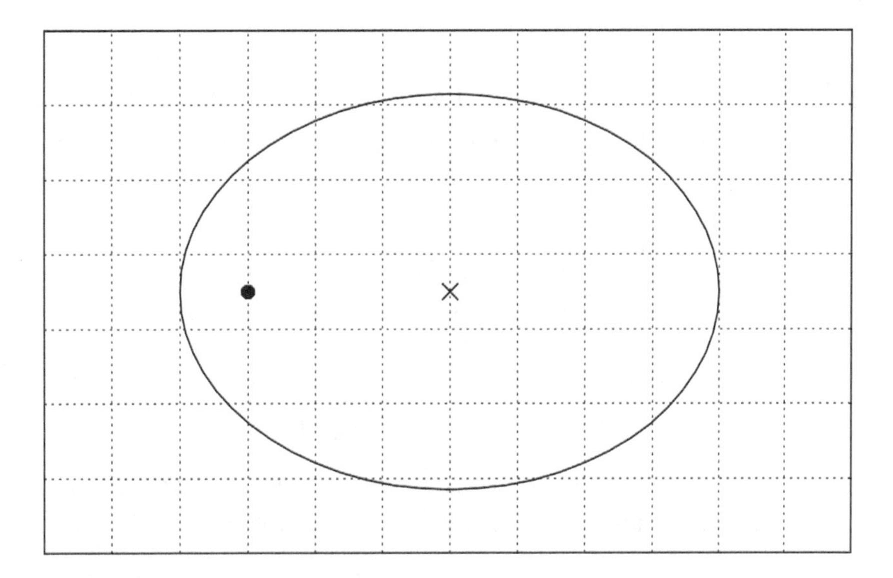

Figure 1: An elliptical orbit. The cross shows the center of the ellipse. Each dotted square measures 1 **astronomical unit** (AU) on a side.

1. One focus of the ellipse is shown by a black dot. Draw the Sun at the other focus.
2. How many AU across is the major axis of this ellipse? ____________________
3. The semi-major axis of this ellipse equals how many AU? ____________________
4. What is the eccentricity of this ellipse? ____________________

Kepler's Second Law

An object moves faster in its orbit when it is closer to the Sun, so that the line connecting the object and the Sun sweeps out equal areas in equal times.

Note that Kepler's Three Laws apply to all objects orbiting the Sun, not just planets.

The point in an object's orbit where it is closest to the Sun, at radius r_p, is called **perihelion**. The point where it is farthest from the Sun, at radius r_a, is called **aphelion**. To sweep out equal areas in equal times, the object moves $\frac{r_a}{r_p}$ times faster at perihelion than at aphelion. Perihelion and aphelion occur on the major axis, so $r_a + r_p = 2a$, where a is the semi-major axis.

5. Mark the perihelion of the orbit in Figure 1 with a P and the aphelion with an A.
6. How much faster is the object moving at perihelion than at aphelion? ____________

Kepler's Third Law

For objects orbiting the Sun, the square of the object's orbital period *P*, measured in years, equals the cube of its semi-major axis *a*, measured in astronomical units:

$$P^2 = a^3$$

The **orbital period** (P) of an object is how long it takes to move around its orbit once. The Earth has $P = 1$ Earth year and semi-major axis $a = 1$ astronomical unit (AU) by definition. For Earth, Kepler's Third Law $P^2 = a^3$ gives $1^2 = 1^3$ or $(1 \times 1) = (1 \times 1 \times 1)$, which is correct.

The orbits of five asteroids (labeled A through E) are shown in Figure 2. Some of the orbits are circular and some are elliptical. Refer to Figure 2 when answering the following questions.

7. What is the perihelion of each orbit, in AU? Fill in these values in the table.
8. What is the aphelion of each orbit, in AU? Fill in these values in the table.
9. What is the semi-major axis of each orbit, in AU? Fill in these values in the table.
10. Rank the orbital periods of the asteroids **from shortest to longest**. Exact values are not needed, just the **relative rankings**. Fill in these relative rankings in the table.

As an example of this kind of ranking, the eccentricities of each of the asteroid's orbits have been ranked and entered in the table. Three of the orbits are circular and, therefore, have zero eccentricity. Of the other two orbits, E is clearly more elliptical than D.

11. One of these asteroids is later discovered to be a comet. **Based just on their orbits,** which asteroid is it most likely to be, and why?

____

____

12. (**Optional**) What are the orbital periods of these asteroids, in Earth years?
13. (**Optional**) What are the exact eccentricities of the orbits of D and E?

NAME ______________________________ ID# ______________________________

DATE ______________________________

Asteroid Name:	A	B	C	D	E
Perihelion in AU					
Aphelion in AU					
Semi-Major Axis in AU					
Eccentricity	Zero	Zero	Zero	Low	Medium
Orbital Period Ranking					
Orbital Period (Bonus)					

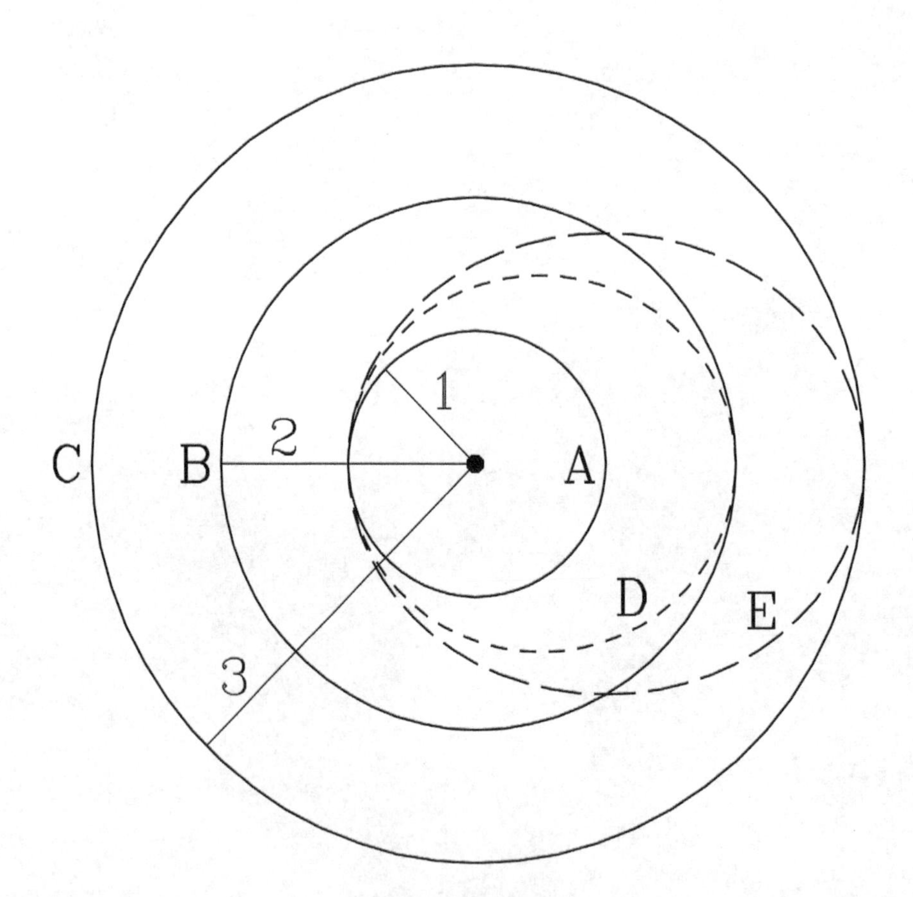

Figure 2: Circular and elliptical orbits of 5 asteroids (A through E) around the Sun (black dot). The numbers show the distances in AU to the circular orbits of asteroids A, B and C.

How Long Are the Days on Mercury and Venus?

NAME ____________________ ID# ____________________ DATE ____________________

A planet's **year** is how long it takes that planet to orbit the Sun.

A planet's **solar day**, usually just called its **day**, is the average time from noon to noon on that planet. In other words, it is the **synodic period** of the Sun as seen from that planet.

The length of a day and the number of days per year is different for each planet. For example, one **Mars day** is 1.027 Earth days long and one **Mars year** is 669 **Mars days** long (687 Earth days).

Let's work out how long the solar days are on the slowly spinning planets Mercury and Venus. To do this, we'll use our knowledge of how long the **sidereal days** are on Mercury and Venus. An object's sidereal day is how long it takes to spin completely around (360°) on its axis. We'll also need to know how long those planets take to orbit the Sun, and in what direction they spin and orbit.

Mercury

Figure 1 shows Mercury at six positions in its orbit around the Sun.

To a reasonable approximation, Mercury orbits the Sun counterclockwise once every 90 Earth days, and Mercury spins 360° counterclockwise on its axis once every 60 Earth days.

That is, the **sidereal day** on Mercury is about 60 Earth days long.

1. On Earth day 0, Mercury is at position A. What Earth day is it when Mercury is at positions B through G? (G is the same as A, but one orbit later.) Write those numbers on Figure 1.
2. At each position, draw in Mercury's **terminator** (the boundary between night and day). Then shade the night side of the planet. This has already been done for you at position F.
3. In both Figure 1 and Figure 2, in position A we have drawn a mountain on Mercury (call it Triangle Mountain).

   You know how long Mercury takes to spin around 360 degrees, so at each position in Figure 2 you can draw in Triangle Mountain.

   When you have filled in Figure 2, copy the location of Triangle mountain at each position onto the drawing of Mercury at the same position in Figure 1.
4. What time is it on Triangle Mountain when Mercury is at positions A, D, and G?

   Use terms like noon, early or late afternoon, sunset, midnight, etc.

   Once you have filled in positions A, D and G, you should be able to fill in the times for positions B, C, E and F. Position A has already been filled in for you:

   A: noon B: __________ C: __________ D: __________ E: __________ F: __________ G: __________

5. From the above, how many Mercury years is it between noon and midnight on Mercury?

   Therefore, how long is it from noon to noon on Mercury?

   Therefore, one **solar** day on Mercury is _____ Mercury years or _____ Earth days long.

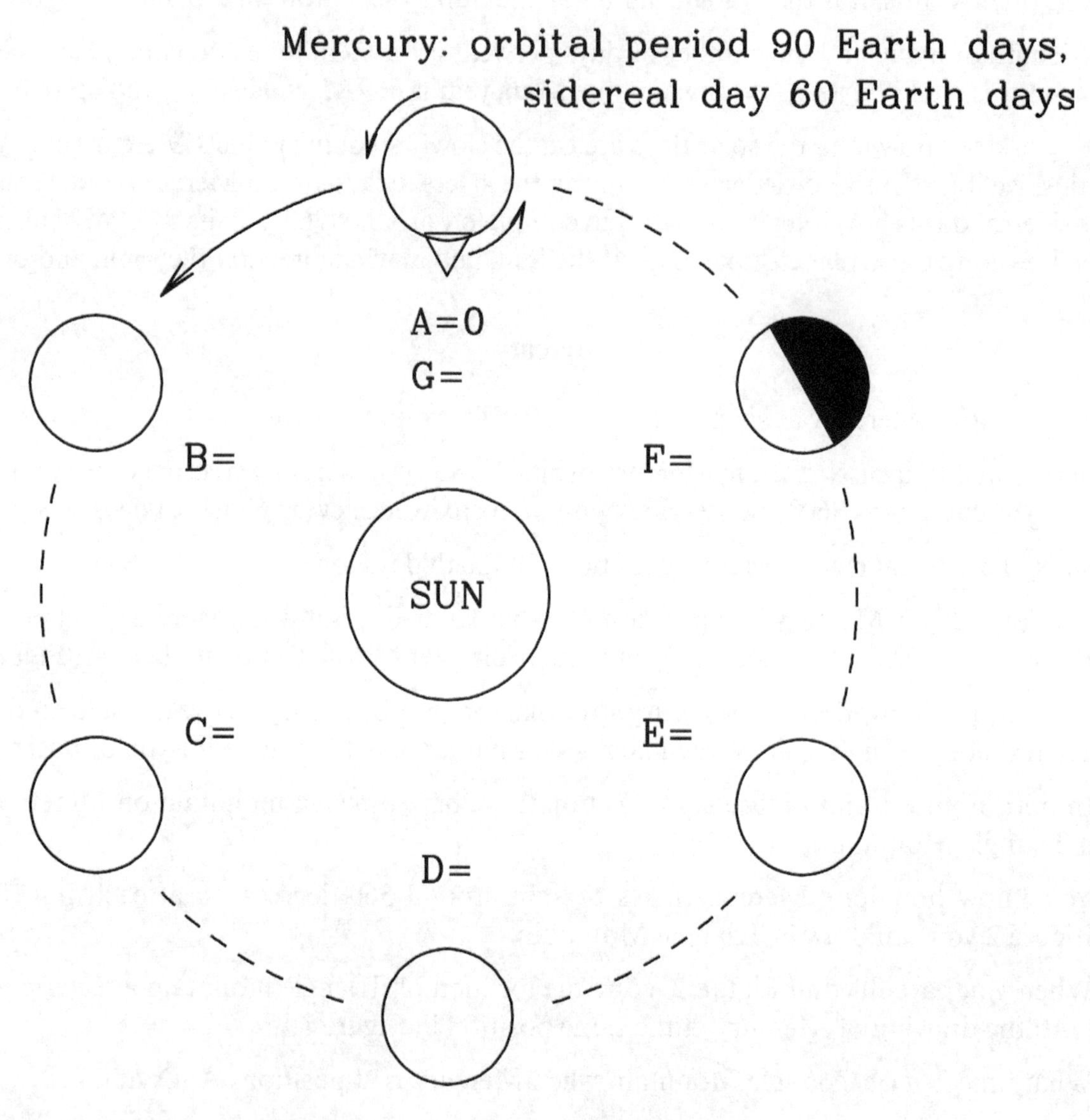

Figure 1: Mercury at six positions in its orbit around the Sun.

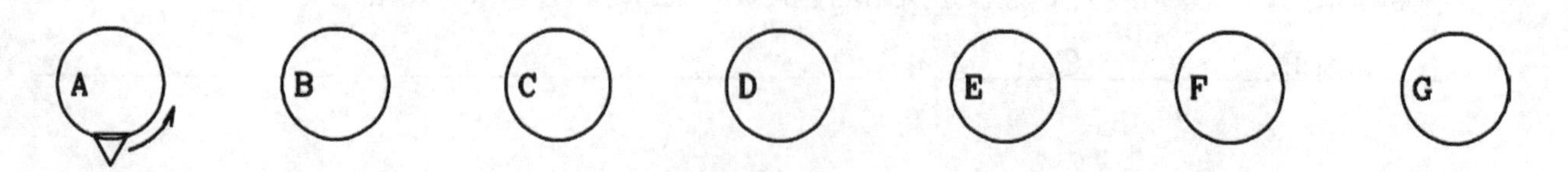

Figure 2: Mercury at the orbital positions of Figure 1. (Use this Figure for Question 3.)

Venus

In Figure 3, Venus is shown in eight positions in its orbit.

Venus orbits the Sun counterclockwise about once every 240 Earth days.

Venus spins 360° CLOCKWISE on its axis about once every 240 Earth days as well.

That is, the **sidereal day** on Venus is (about) 240 Earth days long.

6. At each position in its orbit, draw Venus' terminator, shade in its night side (see position F), and write in the number of the Earth day on which Venus will be at that position.

7. In both Figure 3 and Figure 4, we have drawn a mountain (Triangle Peak) on Venus in position A.

   You know how long Venus takes to spin around 360 degrees, so at each position in Figure 4 you can draw in Triangle Peak.

   When you have filled in Figure 4, copy the location of Triangle Peak at each position onto the drawing of Venus at the same position in Figure 3.

8. What time is it on Triangle Peak at each position in Venus's orbit? (Start with positions A, C, E, G, and I; Position I is the same as position A, but one orbit later.) Position A has already been filled in for you: A: noon

   B:_____ C:_____ D:_____ E:_____ F:_____ G:_____ H:_____ I:_____

$$\frac{1}{P_{\text{sidereal}}} = \frac{1}{P_{\text{y.r.f.orbit}}} + \frac{1}{P_{\text{synodic}}}$$

9. From the above, one **solar** day on Venus is _____ Venus years or _____ Earth days long.

10. Use the formula

    to confirm that the synodic periods of the Sun as seen from Mercury and Venus (the lengths of their solar days) match the results you obtained geometrically in questions 5 and 9.

11. Suppose an extrasolar planet orbits its star clockwise once every 4 days and also spins on it axis clockwise once every 4 days. Can you define the length of a solar day for this planet? If so, how long is a solar day on this planet?

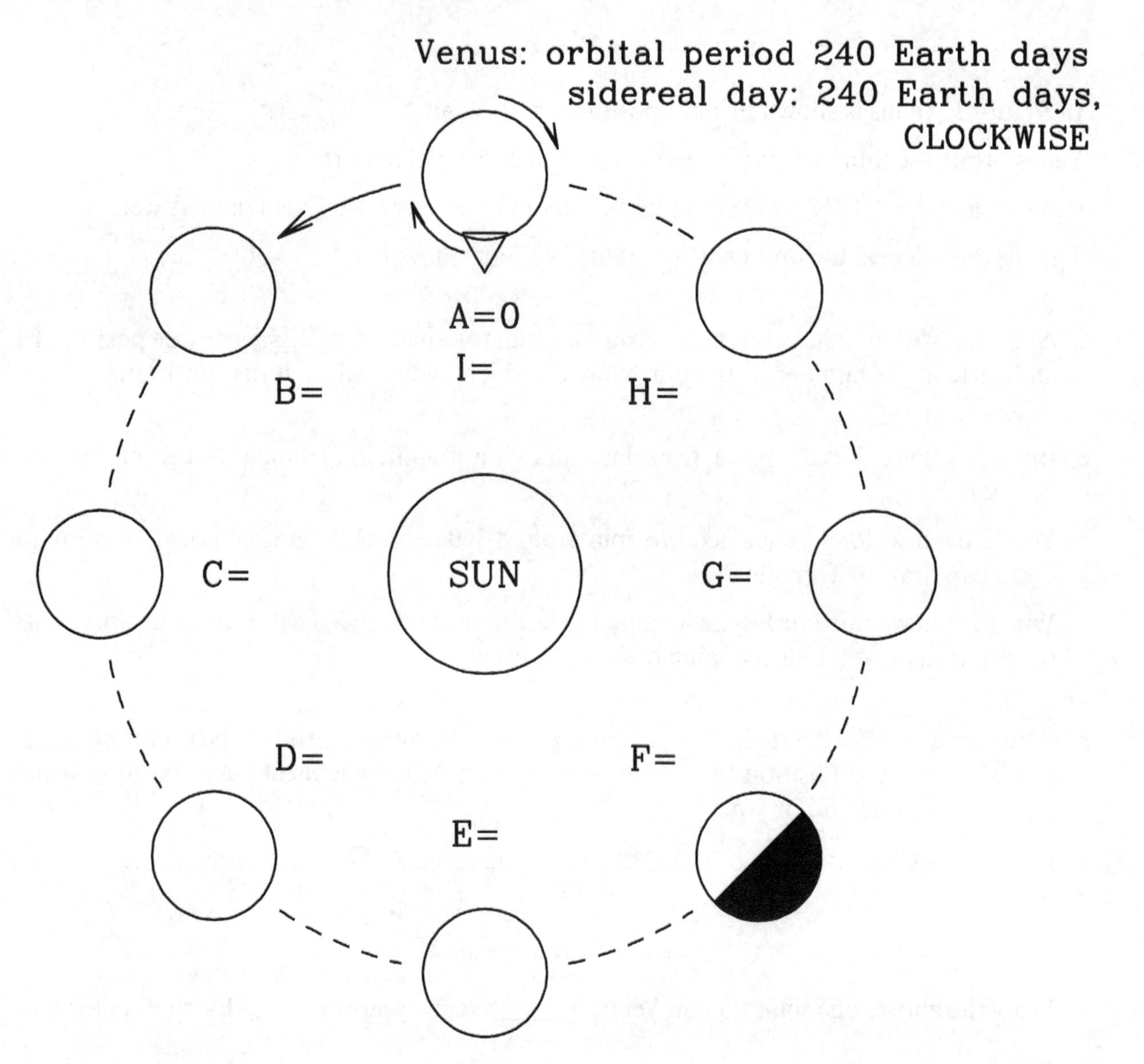

Figure 3: ABOVE: Venus at eight positions in its orbit around the Sun.

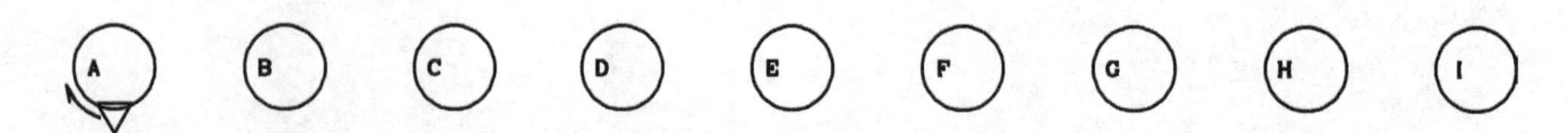

Figure 4: Venus at the orbital positions shown in Figure 3. The location of a mountain on Venus is shown in position A. Draw in the location of the mountain at positions B through I.

Electromagnetic Radiation and Thermal Spectra

NAME ________________ ID# ________________ DATE ________________

Any object can be characterized by a temperature. Temperature measures the average random motion or random vibration of an object's atoms or molecules, which include electrically charged electrons. Random motions or vibrations require accelerations to change directions, and accelerating charged particles emit **electromagnetic (EM) radiation**. Therefore, every **object emits some EM radiation.** EM radiation is often referred to as **light** by astronomers. EM radiation includes not just visible light but also gamma rays. infrared radiation (heat), radio waves, ultraviolet light, and X-rays.

Light consists of "particles" called **photons**. A photon has energy but no mass. Photons of any energy are possible. All photons travel at the speed of light.

A photon can also be thought of as an **electromagnetic wave**, with its **wavelength** (λ) and **frequency** (v) determined by its energy (E). Wavelength is the distance between wave peaks, often measured in **nanometers (nm)**: 1 nm = 1 billionth of a meter $\left(10^{-9}\ \text{m}\right)$. Frequency is the number of times per second that a wave peak passes by you, and is measured in **hertz (Hz)**: 1 Hz = 1 peak per second. (Wavelength and frequency are related by the **speed of light** (c); for all photons, $c = \lambda v$.) The greater the photon energy, the more wave peaks pass by you per second, and the shorter the distance between them:

The frequency (v) of a photon is proportional to its energy: $v \propto E$

The wavelength (λ) of a photon is inversely proportional to its energy: $\lambda \propto \frac{1}{E}$

The above relationship means that if the energy of a photon doubles, its wavelength is halved.

1. Complete this sentence: If photon A has half the energy of photon B, then photon A will have __________ the frequency of photon B and __________ the wavelength of photon B.

Thermal Spectra

The **luminosity** (L) of an object is the light energy per second emitted by that object. It can be measured over all wavelengths (**total luminosity**) or at each wavelength. When you measure an object's luminosity at each wavelength, you have measured that object's **spectrum**. Any sufficiently dense object will emit a particular kind of spectrum known as a **thermal spectrum**. The **shape** of this thermal spectrum depends **only** on the object's temperature, in two ways:

I. A hotter object emits more light per unit area at every wavelength than a cooler object.

II. A hotter object reaches its peak luminosity output per unit area and unit wavelength at a higher frequency (shorter wavelength) than a cooler object. In other words:

The frequency at which the peak luminosity output per unit area and unit wavelength of an object occurs is proportional to the object's temperature: $v_{peak} \propto T$

The wavelength at which the same peak occurs is inversely proportional to T: $\lambda_{peak} \propto \frac{1}{T}$

Figure 1 shows three thermal spectra (**spectra** is the plural of spectrum). The three objects plotted in Figure 1 have temperatures of 5850 K (the temperature of the Sun's surface), 3900 K (two-thirds the temperature of the Sun's surface), and 6435 K (ten percent hotter than the temperature of the Sun's surface).

2. The Sun emits a roughly thermal spectrum (we can ignore the Sun's narrow absorption lines in this exercise). According to Figure 1, at what frequency (ν_{peak}) does the Sun ($T = 5850$ K) emit the highest luminosity per unit area and unit wavelength? ________________

3. The object with $T = 3900$ K has a surface temperature two-thirds of the Sun's. What should its ν_{peak} be, based on your answer to Question 2? ________________ What ν_{peak} do you measure for it in Figure 1? ________________ Are your answers in reasonable agreement? *(Yes | No)*

4. Visible light occupies the grey shaded region in Figure 1. When an object heats up to a sufficiently high temperature, it glows reddish, then yellowish, then bluish. Therefore, which photons have higher energy: red photons or blue photons? ________________ Based on that, correctly label the ends of the grey shaded region in Figure 1 with "blue" and "red".

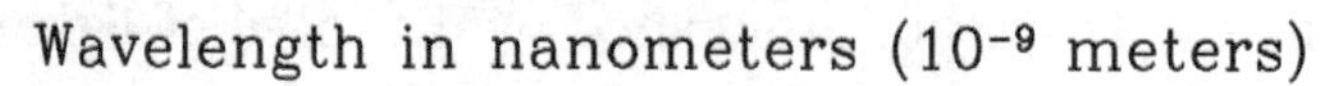

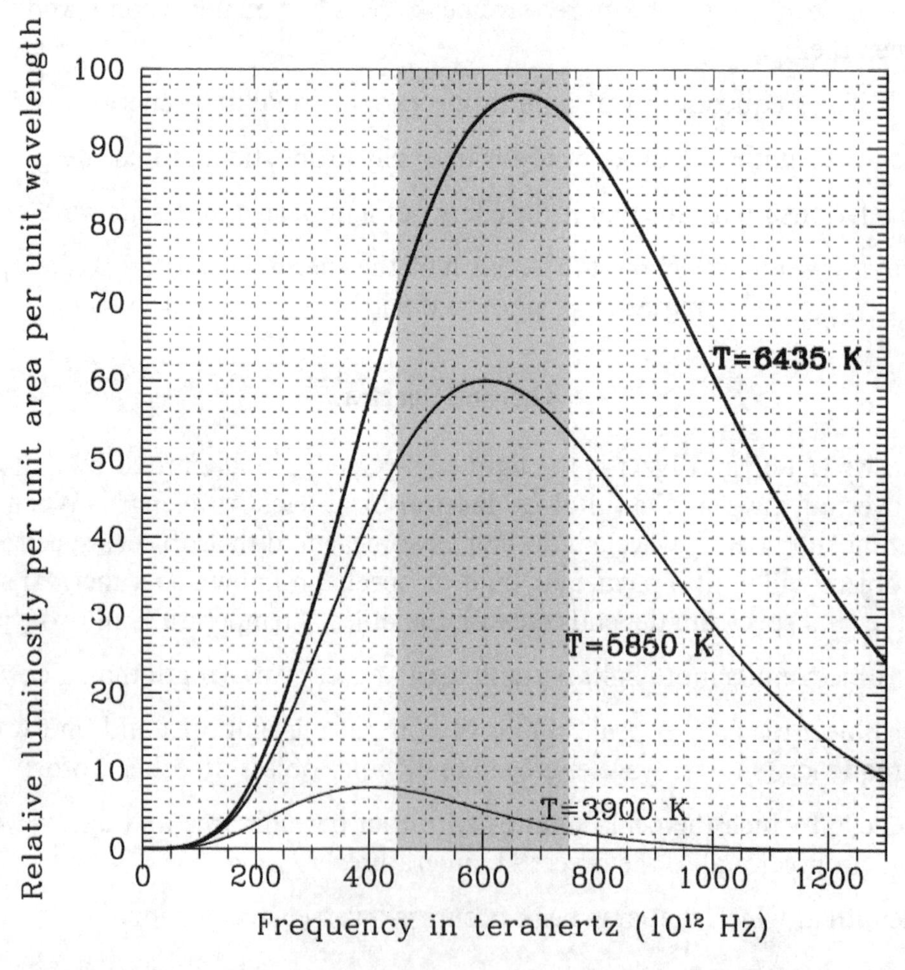

Figure 1: Thermal spectra for three objects, with each object's temperature shown.

Unit 6.2 A Scale Model of The Solar System

Objective

To demonstrate the various sizes and orientations of the Solar System's eight planets as well as lay out the groundwork for constructing a scale model Solar System extending out to Neptune

Introduction

The radius of a planet is a measure of a planet's size, from its uppermost solid layer (or the top of its atmosphere for a gas giant without a discernible solid surface) to the center of its core. Most of the planets are very close to spheres, but the rate at which a planet spins along with its composition causes the planets to bulge slightly at their equators. The most striking example is Saturn, with its very, very low density (about half the density of water) and very fast rotation rate (about 18,000 kilometers per hour) causing the planet's equator to bulge out 6000 kilometers more than its north pole to south pole distance.

Another pronounced characteristic of a planet is its axial tilt. The axial tilt of a planet describes the angle that the planet's equator makes to the planet's orbital path. For a planet with no axial tilt, the line of its north-south pole axis would be perpendicular (90°) to its orbital plane. The equator of an un-tilted planet would lie directly on top of the planet's orbital path. For various reasons – some lost to ancient history; several of the planets are substantially tipped and appear to lean toward or away from the Sun during their orbit, causing their north-south pole line to be bowed toward the path of their orbit. Earth's 23.5° tilt causes the rise of the seasons: when the Earth's orbital position leaves its north pole bowed toward the Sun, we experience the warmer summer months in the northern hemisphere; when Earth's pole is tilted away from the Sun, the northern hemisphere receives less light and experiences the cooler winter months.

Scale models are crucial to the world of science, engineering, architecture, and civil planning. A scale model is a perfectly proportioned miniature model of a much larger object. Before a building is constructed on campus, before the frame of a concept airplane is built, or before a car rolls off the assembly line, a physically smaller but proportionally identical model is created for testing and experimental purposes. The sizes and distances of Solar System objects make their scale difficult to grasp. Given the titanic size and scale of the Solar System, creating a properly scaled model is difficult. Either the planets are large and easily visible and the distances between the model planets are many miles across or the model is compact and small but the planets are microscopic.

However, in the same way as an architect may build a perfectly scaled miniature of a building to demonstrate what the structure will eventually look like, reproducing a scale model of the Solar System will give you insight into both the proportional size of the planets and their properly proportional distances. A scale model takes a size measurement and shrinks it down for all sizes and distances, equally. Unlike an elementary school science fair, where small Styrofoam balls all fit on a small table lined up with one another, you will quickly realize that even a scale model with a small scale can become incredibly large very quickly.

To calculate how large a scaled model must be, there is a need to relate real-world distances with scale model distances. In the case of a scale model, a planet with a diameter many thousands of kilometers wide may be shrunken down to just a few centimeters.

Equations and Constants

Equation	Expression	Variables
Scale Factor	$Scale\ Factor = \dfrac{Real}{Ruler}$	*Scale Factor*: a conversion factor which bridges actual and scale model sizes *Real*: a real-world distance or size,such as kilometers or AU *Ruler*: a distance or size used in a scale model,such as meters or centimeters

Illustrations

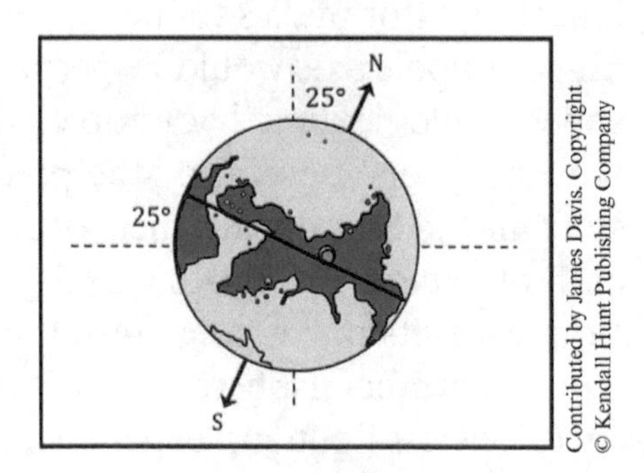

Figure 1. A schematic of the axial tilt of Mars. The dark black line represents Mars's equator with the N-S line representing the planet's polar axis, about which the planet rotates. The dashed horizontal line is the orbit of Mars, the route that it travels around the Sun, and it points directly at the Sun. The dotted vertical line is perpendicular to Mars's orbit and represents the location of Mars's north-south pole line if the planet had zero degrees of axial tilt. Instead, Mars is tilted 25° so its equator rises 25° above its orbital path and its axis likewise leans 25° away from the perpendicular line. The north-south spin axis and the equator are always perpendicular (i.e., they make a 90° angle to one another).

Procedure

The first part of the lab involves drawing accurate representations of the eight planets, including their proper scaled diameters and axial tilts (i.e., their tilt measured with respect to the planetary orbital plane). Earth is 6,370 kilometers in radius, the distance from the core to the surface, with a diameter or 12,740 kilometers. In this model, 6,370 kilometers is set equal to 1.00 centimeter on paper, and all of the planets are drawn relative to this smaller scale. Your ScaleFactor, therefore, would be 6,370 km/cm. You may choose to use a different scale, however.

Using the Scale Factor equation and the actual radii of the eight planets and the Sun, calculate the scale model radius of the given objects, in centimeters. Draw these planets on Worksheets #3 through #7 using the steps listed below. On Worksheet #2, you will properly scale the distance between the miniature planets. Using the same scale factor as on Worksheet #1, convert the Sun-to-planet distances from kilometers into centimeters, using the same process as done on Worksheet #1. Since the numbers will be very large, convert the distances into meters. Using a map, computer program, or a list of locations and distances, choose a location for the center of the Solar System (the Sun). With the scale model distances, build a scale model of the Solar System by determining where the planets should be located (by naming a landmark or intersection that given distance from the Sun). To properly draw the size and orientation of the planets, follow the steps below and refer to the figures:

Step 1: On Worksheets #3 through #7, the orbital plane of each planet has been drawn and labeled. Place a very small **x** mark or dot on the dashed Orbit Line to mark the exact physical center of the planet. Using a ruler, draw a line from the **x** along the orbit with a length equal to the planet's scaled radius in centimeters from Worksheet #1.

Step 2: Using your compass, place the pencil end on the end of your radius line and the metal compass tip on the marked planetary center. Draw a smooth circle representing the planet's circumference. It is usually easier to actually hold the compass still and turn the paper, rather than trying to spin the compass itself around. The finished circle represents the full disk of each planet.

Step 3: Each of the planets has a distinct axial tilt. Using your protractor, place the center mark on the marked planetary center and place a small dot at the angle measurement corresponding to the planet's tilt. Draw a line across the planet's midsection. This is the geometrical equator of the planet and has a length equal to the scaled planet's diameter.

Step 4: Now line the protractor up on the equator – laid on the center mark again – and mark a small point at 90°. Draw a second straight line completely through the planet, perpendicular to the axis. This represents the north-south axis of the planet. Label the north pole with an N and the south polewith an S.

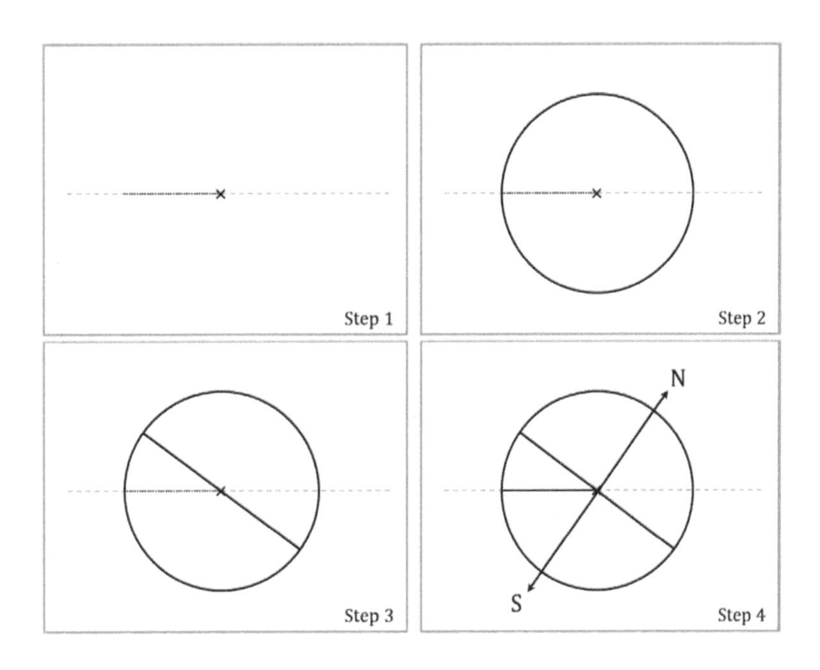

NAME ______________________ ID ______________________

DUE DATE __________ LAB INSTRUCTOR __________ SECTION __________

(example)

$\div \frac{6050}{6370} = 0.95\text{cm}$

Worksheet # 1

Scale Factor Used (in km/cm)
6,370

Star	Radius (in km)	Scaled Radius (in cm)	Axial Tilt (in degrees)
Sun	695,500		0°

Planet	Radius (in km)	Scaled Radius (in cm)	Axial Tilt (in degrees)
Mercury	2,440	0.38cm	0°
Venus	6,050	0.95cm	177°
Earth	6,370	1cm	23.5°
Mars	3,390	0.53cm	25°
Jupiter	69,900	10.97cm	3°
Saturn	58,200	9.14cm	27°
Uranus	25,400	3.98cm	98°
Neptune	24,600	3.86cm	29°

Note: The axial tilt measures the angle between the plane of a planet's orbital plane and its equator.

NAME ______________ ID ______________

DUE DATE ________ LAB INSTRUCTOR ________ SECTION ________

Worksheet # 2

Planet	Distance (in km)	Distance (in AU)	Scaled Distance (in m)	Placement in Model
Mercury	57,910,000	0.387		
Venus	108,208,000	0.723		
Earth	149,598,000	1.00	234.2m	
Mars	227,939,000	1.52		
Jupiter	778,547,000	5.20		
Saturn	1,433,449,000	9.58		
Uranus	2,870,671,000	19.2		
Neptune	4,498,542,000	30.1		

6,370 Km/cm

149,598,000
÷ 6,370

= 23,484,7724 cm

100 / to get metres

= 234.2 m

NAME ______________________ ID ______________________

DUE DATE __________ LAB INSTRUCTOR __________ SECTION __________

Worksheet # 3

The Inferior Planets: Mercury and Venus

Orbit

Line

Orbit

Line

NAME ______________________________ ID ______________________________

DUE DATE __________ LAB INSTRUCTOR ___________ SECTION __________

Worksheet # 4

The Outer Terrestrial Planets: Earth and Mars

← axis tilt estimated

Orbit
Line

← til more than earth

Orbit
Line

NAME ______________________________ ID ________________________________

DUE DATE __________ LAB INSTRUCTOR ____________ SECTION __________

Worksheet # 5

The Gas Giant: Jupiter

Orbit

Line

NAME ______________________ ID ______________________

DUE DATE __________ LAB INSTRUCTOR __________ SECTION __________

Worksheet # 6

The Gas Giant: Saturn

Orbit

Line

NAME ______________________ ID ______________________

DUE DATE __________ LAB INSTRUCTOR __________ SECTION __________

Worksheet # 7

The Ice Giants: Uranus and Neptune

Orbit
Line

Orbit
Line

NAME ______________________ ID ______________________

DUE DATE __________ LAB INSTRUCTOR __________ SECTION __________

Worksheet # 8

Postlab Questions

For each of the following questions, include all work, equations, and proper units.

1. The severity of seasonal temperature swings – summer to winter – is based on the tilt of a planet's axis toward (in summer) and away from (in winter) the Sun. The larger the tilt, i.e., the closer the poles lie to the orbital plane, the more extreme seasons are expected. Which *single planet* in the Solar System would have the most severe seasons in this case? Which *planets* would have the least severe seasons?

2. Saturn's brightest sets of rings, from one edge to the other, are 280,000 km wide. How many centimeters across would you draw the rings in your scale model?

3. The closest star to the Earth besides the Sun is the red dwarf star Proxima Centauri, located at a distance of 4.26 light years (269,000 AU). Using the same scale as used for the Solar System, how far away should Proxima Centauri be placed in a scale model (include units).

4. If 1 mile = 1610 meters, how far away should Proxima Centauri be from the scale model Sun (include units)?

5. The furthest apart one can place two objects on planet Earth is about 8,000 miles (two opposite sides of the planet). What is the immediate problem you see with building a scale model encompassing the Earth, planets, and Proxima Centauri?

Unit 6.5 Kuiper Belt Objects

Objective

To learn about the properties of the Kuiper Belt Objects orbiting beyond Neptune, including their physical sizes, orbital eccentricities, and orbital inclinations

Introduction

Following the discovery of Uranus in 1781 and Neptune in 1846, astronomers searched for signs of other planets lurking in far reaches of the Solar System on long-period orbits for centuries. It has not been until Clyde Tombaugh, working at the Lowell Observatory (Flagstaff, Arizona), spotted a small pinpoint of moving light on a photographic plate in 1930. A new planet exterior to the orbit of Neptune had been discovered. The astronomer Gerard Kuiper theorized that the formation of the Solar System would leave a ring of icy, leftover material in the far reaches of the Solar System, from just beyond Neptune's orbit (~30 AU) out to about 55 AU. Like the asteroid belt between Mars and Jupiter, the Kuiper Belt would be a wide disk of planetary debris slowly orbiting the Sun. It was from this location (aside from the Oort Cloud) that the Solar System's comets would originate: small chunks of ice and rock that vaporize spectacularly as they approach the Sun. Kuiper theorized that many billions of tiny icy fragments may orbit out in that vast region, with a comet created every time a collision or gravitational nudge disturbed the orbit and caused it to careen toward the Sun.

The first Kuiper Belt Object (KBO) fitting this description was 1992 QB_1, discovered in 1992, showing for the first time that there were objects in the outskirt of the Solar System, inhabiting the same region of space as Pluto. In 2005, the object 2003 UB_{313} was identified to be about the same size as but more massive than Pluto, eventually dubbed Eris. Eris joined other substantial KBOs, like Makemake and Haumea. Taken together, these objects showed that Pluto composed only a tiny fraction of the Kuiper Belt's mass, representing a less dominant object than any of the Solar System's eight planets and more a primitive, unfinished remnant of planetary formation. Pluto was subsequently reclassified from planet to dwarf planet.

Today, over 1000 KBOs have been confirmed, with enough observations to allow astronomers to plot out and predict their orbits. The simple Kuiper Belt turned out to be more complex than originally thought, consisting of different, distinct families of KBOs. The objects called Plutinos (like, nor surprisingly, Pluto) show effects of being strongly influenced by Neptune's gravity. Plutinos have perihelion distances which are very close to Neptune's semi-major axis distance of 30.1 AU. In addition, Plutino orbits synchronize with Neptune's orbit, with lengths that are 1.50, 2.00, or 2.50 times the length of Neptune's 164.8 year orbit (i.e., orbits of 248 years, 330 years, or 412 years, respectively).

A second class of KBO, called Cubewanos (like the namesake object QB_1) have orbits that are unperturbed and unaffected by Neptune. Their eccentricities are very small, usually less than 0.1, and have perihelion distances that are nowhere near Neptune's 30.1 AU semi-major axis. Finally, the Kuiper Belt is populated by some objects showing a history of having been badly disturbed into elongated, chaotic orbits by interactions with Neptune or some other distant object. These bodies are called Scattered Disk Objects (SDOs). Scattered disk KBOs are marked by high inclinations, with tilted orbits that lie high above or below the orbits of the eight planets, along with aphelion and perihelion distances that vary greatly and never get very close to Neptune's orbit (with eccentricities larger than 0.25). The scattered disk serves as evidence that the early Solar System was a more disturbed, chaotic place, with these KBOs serving as reminders of the past upheaval.

Equations and Constants

Equation	Expression	Variables
Aphelion	$Q = a(1+e)$	Q: the aphelion distance a: the semi-major axis of the orbit e: the eccentricity of the orbit
Perihelion	$q = a(1-e)$	q: the perihelion distance a: the semi-major axis of the orbit e: the eccentricity of the orbit
Kepler's Third Law	$a^3 = P^2$	a: the semi-major axis of the orbit in AU P: the period of the orbit in years

Procedure

For Worksheet #1, you will identify the largest KBOs by their size. Refer to Datasheet #1 as well as Table 3 in the Appendix for information. The largest circle (marked by a dotted line) represents the circumference of Mercury, the Solar System's smallest planet. The other circles (labeled 1 through 11) represent the 11 largest known KBOs (including Pluto's largest moon, Charon). Despite its small size, Mercury's diameter of 4880 kilometers is over twice the diameter of the largest KBO, Pluto. The drawings of the KBOs' circumferences are scaled so that 1.0 millimeter in the drawing is equal to 36 kilometers of actual length. Because many KBOs have remarkably similar diameters, which appear almost identical in the drawings, several of the objects have already been identified on Worksheet #1. By measurement and elimination, determine which circles correspond to the physical sizes of the remaining 7 KBOs.

Also on Worksheet #1, choose 5 KBOs (excluding Sedna and Pluto's moon Charon). Enter their names. From the Appendix, enter the semi-major axis of the KBO and its eccentricity and calculate both its aphelion distance (in AU) and its perihelion distance (in AU).

Part of Pluto's reclassification from planet to dwarf planet was based on its inability to "clear its neighborhood" of other KBOs. This means that Pluto's orbital region overlaps with the orbital regions of numerous other KBOs. On Worksheet #2 there are five columns, each with a space below to record the name of a KBO from Worksheet #1 and with an axis labeled with distances running from 0 AU (the location of the Sun) to 100 AU (far beyond the outer edge of the Kuiper Belt). The shaded horizontal boxes represent the aphelion and perihelion distances of the eight Solar System's planets, showing the domain dominated by them. Note that Uranus is the most eccentric planet, with a widely spaced perihelion and aphelion distance. Also note that none of the eight planets' domains overlap (or even draw close to one another). Using your KBOs and their calculated aphelion and perihelion distances from Worksheet #1, plot a pair of lines across the single column to represent the perihelion and aphelion distance of each KBO. Shade in the space between those lines. This represents the "neighborhood" of the KBO. For a planet, that neighborhood would be free of any overlapping neighborhoods from the other columns. This will clearly not be the case for the KBOs, as you may notice significant regions of overlap.

For Worksheet #3, use your five chosen KBOs again and look up their orbital inclinations (the tilt of their orbit relative to the plane of the Earth's orbit around the Sun). Mark with a small dot the angular position of those KBO's orbits, connect the dot to the location of the Sun with a dashed line (representing the plane of the planet's orbit) and label the point with the KBO's name.

Datasheet #1

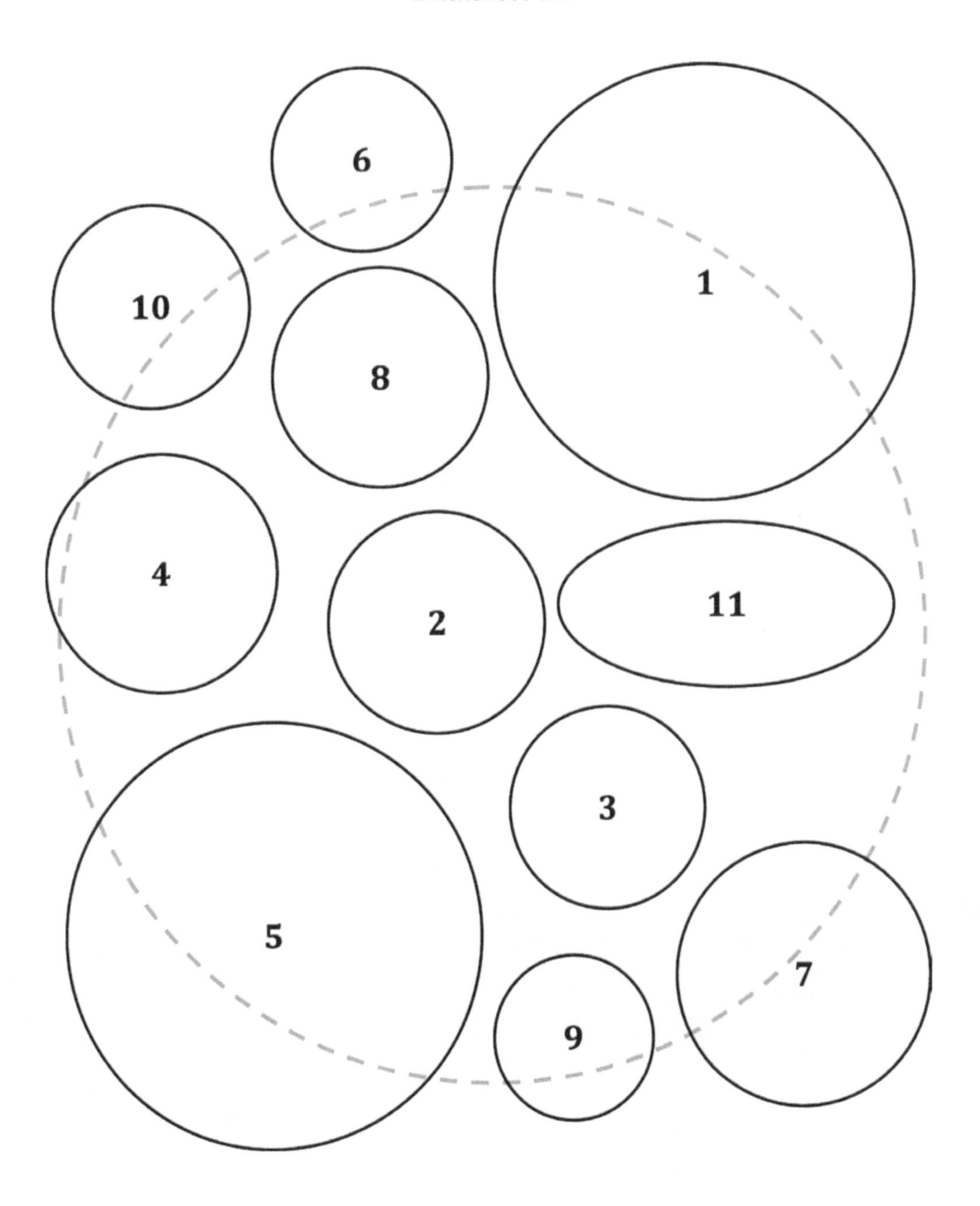

NAME ______________________________ ID ______________________________

DUE DATE __________ LAB INSTRUCTOR ______________ SECTION __________

Worksheet # 1

KBO Sizes

Object	KBO Name	Measured Diameter (in mm)	Calculated Diameter (in km)
1	Pluto	66	2370
2	Charon	34	1200
3	Orcus	31	1110
4	2007 OR_{10}	36	1290
5			
6			
7			
8			
9			
10			
11			

KBO Orbits

KBO Name	Semi-Major Axis (in AU)	Eccentricity	Perihelion (in AU)	Aphelion (in AU)

Data obtained from:

http://www2.ess.ucla.edu/~jewitt/kb/big_kbo.html

NAME ______________________ ID ______________________

DUE DATE __________ LAB INSTRUCTOR ______________ SECTION __________

Worksheet # 2

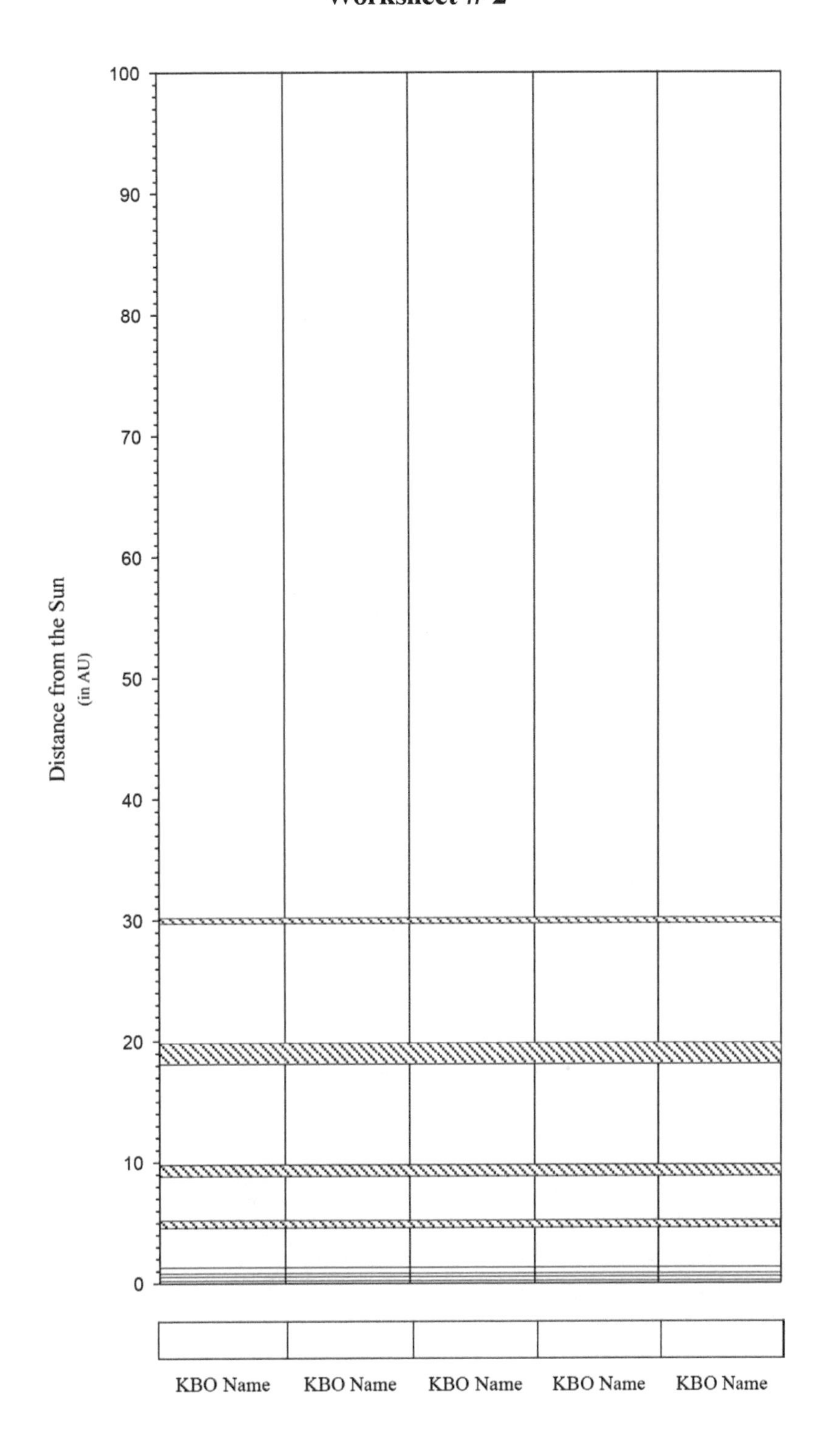

NAME ______________________________ ID ______________________________

DUE DATE ________ LAB INSTRUCTOR ______________ SECTION __________

Worksheet # 3

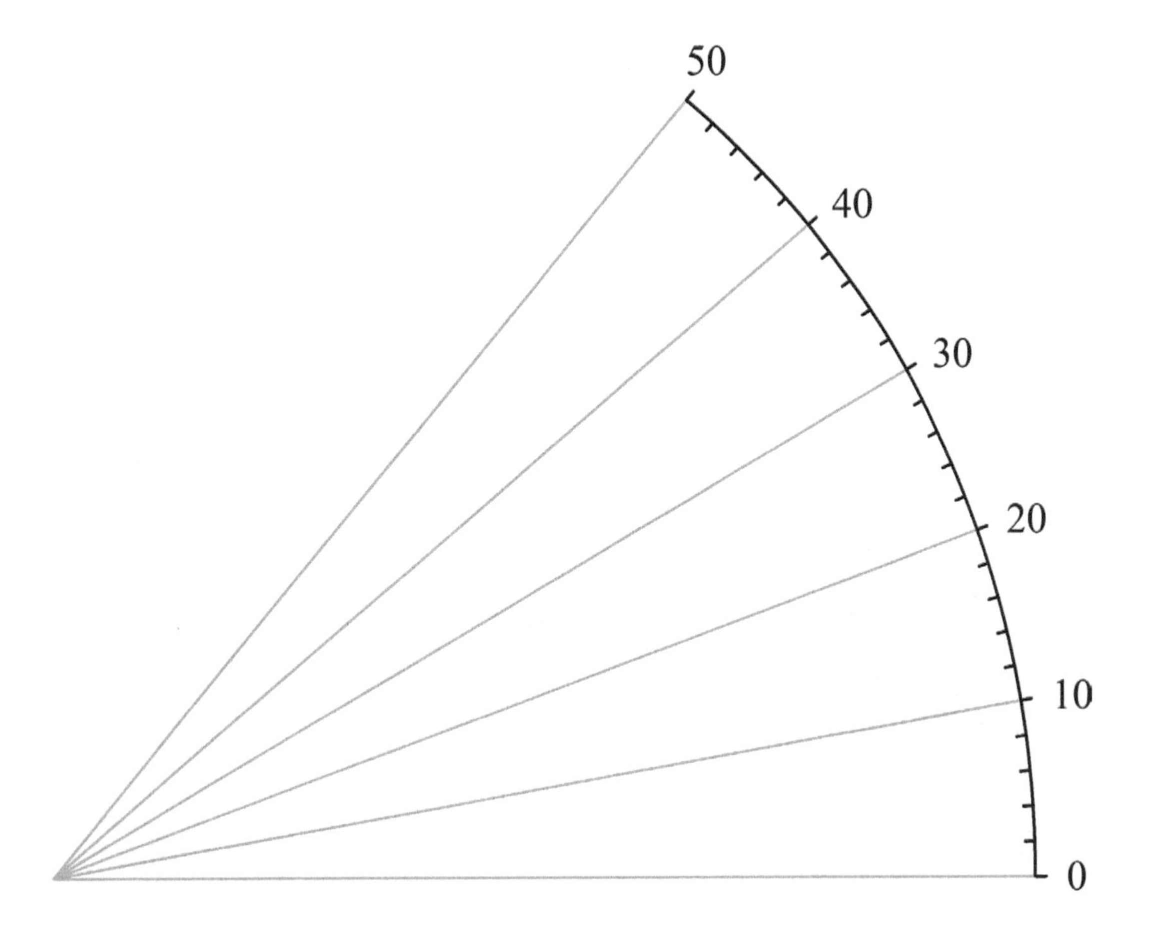

NAME ______________________ ID ______________________

DUE DATE ________ LAB INSTRUCTOR ____________ SECTION ________

Worksheet # 4

KBOs fall into three classes, as covered in the Introduction: cubewanos, plutinos, and SDOs. Using the data for the KBOs from the Appendix Table 3, answer the following questions about the nature of some KBOs. Calculate the necessary values of perihelion, period, and period-ratio or state pertinent values given in the KBO data table, such as inclination or eccentricity. Points are awarded on thoroughness of answers, so be as complete as possible in your replies.

1. Referring back to the orbital inclinations of your KBOs as selected, why even if two KBOs have overlapping "neighborhoods" or share an orbit, they will be very unlikely to crash into one another?

2. What family of KBOs does Orcus belong to? Is it a Cubewano, a Plutino, or an SDO? Show calculations or measurements that lead to this conclusion and explain your answer.

3. What family of KBOs does Quaoar belong to? Is it a Cubewano, a Plutino, or an SDO? Show calculations or measurements that lead to this conclusion and explain your answer.

4. What is Eris's orbital period, in years?

5. Today, Sedna is about as close to the Sun as its orbit allows it to be, which allowed astronomers to detect it before its long orbit takes it far out of the Kuiper Belt and makes it invisible to most telescopes. What is Sedna's period, in years?

6. In 2014, astronomers discovered 22 new KBOs, among those the following three: 2014 UF_{224}, with a period of 331.1 years; 2014 TT_{85}, with a period of 280.8 years; 2014 QM_{441}, with a period of 315.2 years. Given that information and knowing that Neptune's period is 164.8 years, which of those objects – if any – are likely part of the Plutino family of KBOs? Show your work that lead you to your conclusion.

Age Dating Through Radioactive Decay

NAME ______________________________ ID# ______________________________

DATE ______________________________

An atomic nucleus consists of particles known as protons and neutrons. Every nucleus of a given **element** has the same number of protons: hydrogen has one proton, helium has two protons, etc. Nuclei of the same element with different numbers of neutrons are known as **isotopes**. For example, deuterium (1 proton + 1 neutron) and tritium (1 proton + 2 neutrons) are isotopes of hydrogen.

If an atomic nucleus has too few or too many neutrons relative to its number of protons, it will be **unstable** and will decay into a different nucleus; this is called **radioactive decay**. A nucleus with too many neutrons can decay when one neutron turns into a proton and an electron. For example, tritium decays into helium-3 (2 protons and 1 neutron) and an electron.

Each unstable isotope has a 50% chance of decaying within a time span known as that isotope's **half-life**.

A single unstable isotope can decay at any time, but in a large sample of nuclei of that isotope, you will find that 50% of the nuclei will decay during the first half-life you study, 50% of the remaining nuclei will decay during the next half-life you study, 50% of the remaining nuclei will decay during the next half-life after that, etc.

Consider a specific (but fictonal) example. Element Tr (for Triangle) decays to element Fc (filled circle). In Figure 1, 16 newly created nuclei of element Tr are shown at time zero on the left. (The nuclei might be newly created in a supernova, for example.) Each Tr nucleus is shown as a triangle inside a circle. In the same Figure, those same 16 nuclei are also shown at four **equally spaced** later times. Use Figure 1 to answer the following questions.

1. At time 1, how many nuclei of element Tr are left? ____________
2. What fraction of the original nuclei of element Tr are left at time 1? ____________
3. What do we call the amount of time between time 0 and time 1? ____________
4. Complete Figure 1 by drawing in the correct number of Tr nuclei (circled triangles) and Fc nuclei (filled circles) at times 2, 3 and 4.
5. Fill in the table below with the number of nuclei of element Tr and of element Fc at each time. Then find the ratio of the number of nuclei of element Fc to the number of nuclei of element Tr. That ratio is what is used to estimate the age of a sample of material containing an element and the radioactive decay product of that element. Fill in that column of the table.

Time	# of Tr nuclei	# of Fc nuclei	Ratio Fc/Tr (Product Isotope/Parent Isotope)
0	16	0	0
1			
2			
3			
4			

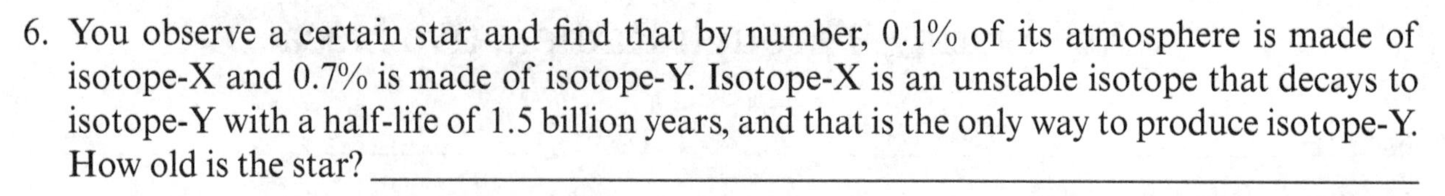

6. You observe a certain star and find that by number, 0.1% of its atmosphere is made of isotope-X and 0.7% is made of isotope-Y. Isotope-X is an unstable isotope that decays to isotope-Y with a half-life of 1.5 billion years, and that is the only way to produce isotope-Y. How old is the star? ____________________

The next two questions are about carbon dating. Trees continuously absorb the unstable isotope carbon-14 from the atmosphere. When a tree is chopped down, no more carbon-14 is absorbed, and the existing carbon-14 decays to nitrogen-14 with a half-life of about 6000 years.

7. A sample of a piece of burnt wood from a cave in Freedonia contains 14 to 15 parts per million of nitrogen-14 and 1 to 2 parts per million of carbon-14. What is the range of ages possible for when the tree was chopped down?

____

8. A new, more sensitive analysis is applied to another small sample of the same piece of wood. It is found to contain 15,000 to 15,015 parts per billion of nitrogen-14 and 1,000 to 1,001 parts per billion of carbon-14. What is the revised range of ages possible for this charcoal?

____

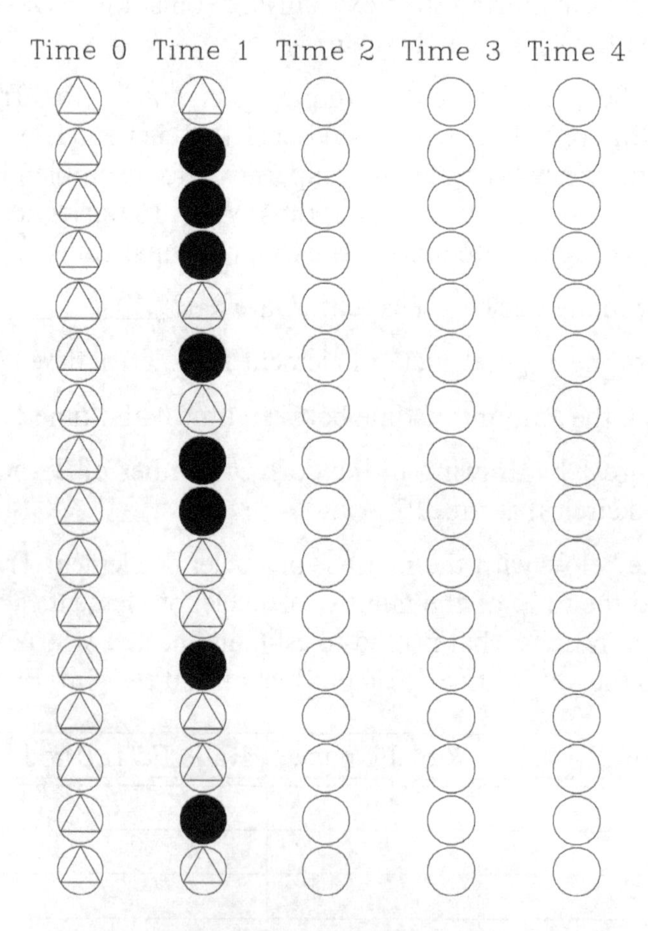

Figure 1: Sixteen nuclei of element Tr are shown at time zero. At time 1, some nuclei have radioactively decayed to element Fc (filled circles). The situations at later times are left blank.

Tidal Forces and the Roche Limit

NAME ______________________________ ID# ______________________________

DATE ______________________________

Imagine a solar system with a planet orbited by two moons (Figure 1). The moons are identical except that one orbits farther from the planet than the other. The center of moon 1 is located 14 moon radii from the center of the planet, and the center of moon 2 is located 9 moon radii from it. Each moon has mass m=20, the planet has mass M=10,000, and the force of the planet's gravity *on an object of mass m=1* (for Newton's G=1) has been calculated at each distance in Figure 1. For example, the force of gravity on an object of mass m=1 at the center of moon 1, a distance d=14 from the center of the planet, is $F = \frac{G \times M \times m}{d \times d} = \frac{1 \times 10{,}000 \times 1}{14 \times 14} = 51$.

1. The orbital motion of each moon balances the force of the planet's gravity as measured at the *center* of each moon, which is why the moons stay in orbit instead of falling toward the planet. However, this orbital motion does not exactly balance the force of the planet's gravity at the *surface* of each moon. Those unbalanced forces are called **tidal forces**, and we can use arrows to see their effects. *Fill in the blank lines next to each moon in Figure 1 by summing the two numbers to the left of each line. The resulting numbers are the tidal forces at points A and B on moon 1 and at points C and D on moon 2.* The calculations at the center of each moon have been done for you: both moon 1 and moon 2 experience zero tidal force at their centers.

2. Let's compare those tidal forces to the force of gravity at the surface of each moon. In question 1 you calculated tidal forces at points A, B, C, and D. *Now draw arrows of those lengths at A, B, C, and D, pointing in the appropriate direction for a negative or positive force.* **Useful information:** • Each moon's force of gravity (on an object of mass m=1) is shown by the arrows already drawn at points A, B, C, and D, pointing toward the center of each moon. • The length of each arrow represents 20 force units. • When a force points toward the planet, it is negative. • When a force points away from the planet, it is positive.

3. In what direction relative to the *surface* of moon 1 is the tidal force trying to move objects at point A? ____________ At point B? ____________________

4. In what direction relative to the *surface* of moon 2 is the tidal force trying to move objects at point C? ____________________ At point D? ________________

5. In Figure 1, a longer arrow means a stronger force. Is the tidal force stronger than the moon's gravity at point A on moon 1? __________ ...at point B on moon 1? __________ ...at point C on moon 2? __________ ...at point D on moon 2? __________

6. At any point where you answered no to question 5, the moon's surface will just be pulled a bit in the direction of the tidal force. *If you answered no for moon 1 or moon 2, draw the resulting shape of the surface of the moon(s)* under "Tidally distorted moons and planet" in Figure 1. The dashed circles show the original shapes of each object, and the sides of each tidally distorted object have been drawn in for you as short solid lines. Which is most

distorted by tidal forces: moon 1, moon 2, or the planet? ________________ Which is least distorted? ______________________

7. At any point where you answered yes to question 5, objects on the surface of the moon will be pulled off into space by the tidal force. If the tidal force is much stronger than gravity, then the surface of the moon itself will be pulled off, and the moon will eventually be pulled apart. *Mark any point on moon 1 or moon 2 where the tidal force is larger than the moon's gravity.*
8. We define the **Roche limit** as the orbital radius at which the tidal force on both sides of an object is greater than the object's gravity. At what distance from the planet is the Roche limit for these moons located? ________________ *Draw it in on Figure 1!*

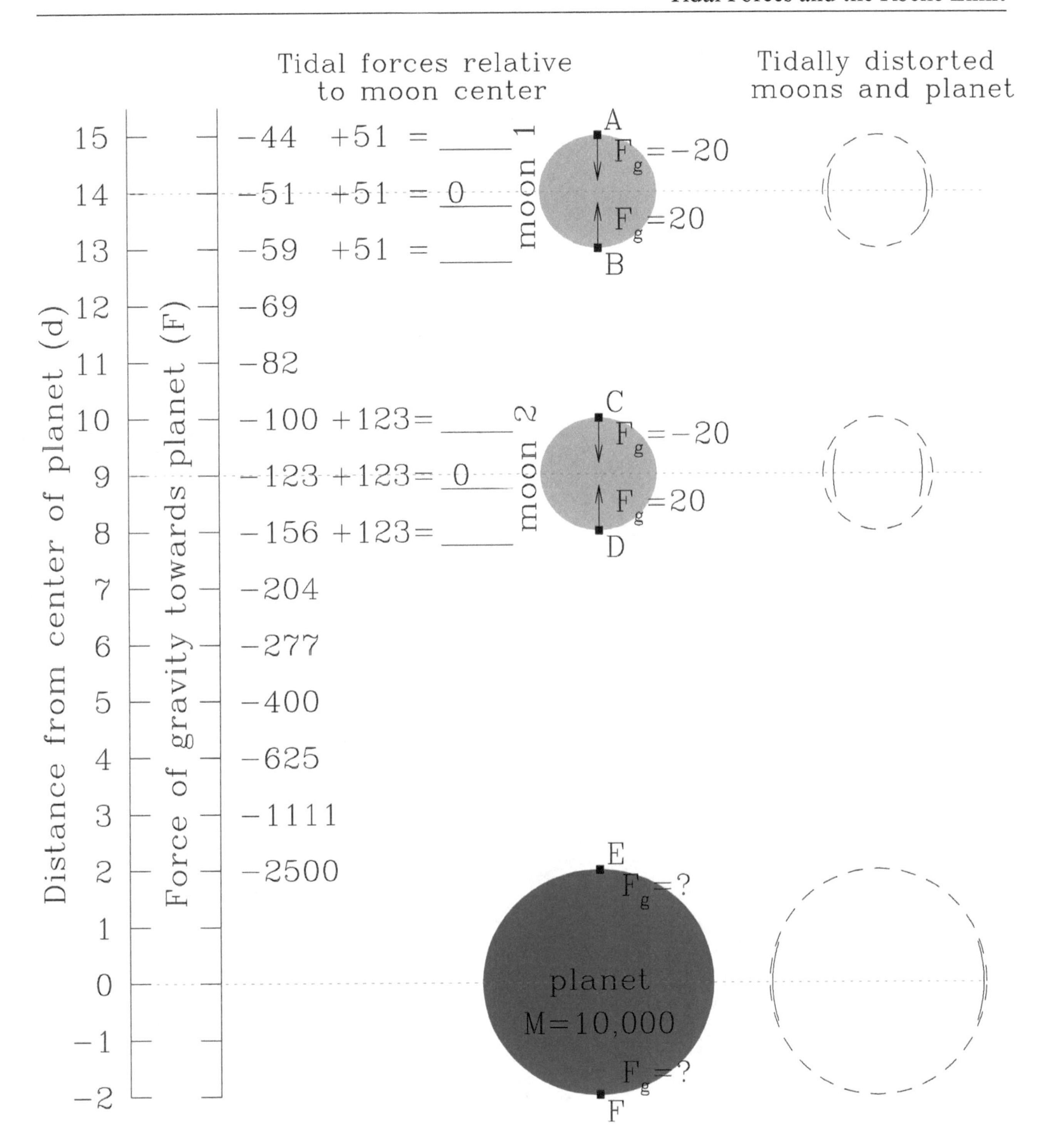

Figure 1: A hypothetical planet with two moons in a distant solar system.

Stellar Masses with Newton's Version of Kepler's Third Law

NAME ______________________ ID# ______________________

DATE ______________________

The centre of mass of two orbiting objects always lies on a line drawn between those objects, and never moves. In a **binary star** system, two stars make elliptical orbits with the centre of mass at one focus. *If two objects have masses* M_1 and M_2 *measured in solar masses* (our Sun has 1 solar mass) and have orbits of semi-major axis a and period *P* around the centre of mass, then Newton's version of Kepler's Third Law is:

$$(\text{Orbital period in Earth years})^2 = \frac{(\text{semi-major axis in AU})^3}{M_1 + M_2} \longrightarrow \boxed{P^2 = \frac{a^3}{M_1 + M_2}} \quad (1)$$

Also, the distances d_1 & d_2 of objects 1 & 2 from their centre of mass are related by $\boxed{\frac{d_1}{d_2} = \frac{m_2}{m_1}}$

where *the distance between the objects is* $\mathbf{d = d_1 + d_2}$. (If the objects were weights connected by a stick, the centre of mass is where you could balance the stick on your finger.)

Let's draw centres of mass and orbits in Figure 1 using Newton's version of Kepler's 3rd law. This has been done for you in the leftmost system; read the figure caption for the details. And in systems I and II, *the dashed line shows the orbit the light grey object would have if the dark grey object had 1000 times the mass of the light grey object.*

1. In system I, objects A and B have the same mass.

   Mark the centre of mass. Assuming circular orbits, draw the orbits of both objects.

2. In system II, object C has three times the mass of object D.

   Mark the centre of mass. Assuming circular orbits, draw the orbits of both objects.

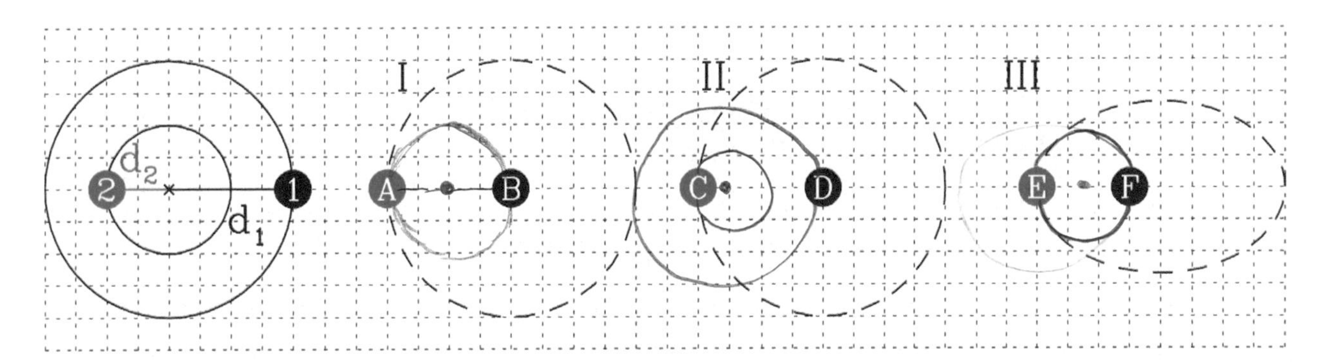

Figure 1: Objects orbiting their centres of mass. In the leftmost system, object 2 has twice the mass of object 1, so its distance from the centre of mass (the ×) is $d_2 = \frac{1}{2}d_1$. *It is always true that the more massive object traces out a smaller orbit than the less massive object does.*

3. In system III, objects E and F have the same mass. The elliptical orbit of object E around the centre of mass is shown. Mark the centre of mass and draw in the orbit of F.

Finding the Masses of Orbiting Objects

Newton's version of Kepler's Third Law enables us to determine the mass of **any** objects orbiting each other (stars, for example) **if** we can determine their orbital period and semi-major axis. *If two objects have masses M_1 and M_2 measured in solar masses* (our Sun has 1 solar mass) and have orbits of orbital period P and semi-major axis a around the centre of mass, then Newton's version of Kepler's Third Law is:

$$(\text{Orbital period in Earth years})^2 = \frac{(\text{semi-major axis in AU})^3}{M_1 + M_2 (\text{in solar masses})} \longrightarrow \boxed{P^2 = \frac{a^3}{M_1 + M_2}} \qquad (2)$$

For elliptical orbits, the separation d changes with time but a is always given by : $a = \frac{d_{min} + d_{max}}{2}$ where d_{min} and d_{max} are the minimum and maximum separation of the objects.

4. On the previous page, if stars A and B in system I take 2 years to orbit each other, how much mass do each of them have?

____

5. On the previous page, if stars C and D in system II take 1 year to orbit each other, how much mass do each of them have?

____

6. On the previous page, if stars E and F in system III take 2 years to orbit each other, how much mass do each of them have?

____

7. Suppose you observe *two identical stars, each with half the mass of our Sun*, which orbit their center of mass with a semi-major axis of 1 AU. How many years does it take them to complete one orbit?

____

8. Suppose you observe two stars (1 and 2) orbit their centre of mass with a semi-major axis of 2 AU, taking 4 years to complete one orbit.

   What is $M_1 + M_2$ for these stars? _______________________

   If the stars are identical, what is M_1 and what is M_2? _______________________

9. You observe two stars and find that one star appears to be stationary on the celestial sphere while the other moves around it in a circular orbit with a semi-major axis of 4 AU, making one orbit every 4 years. What is the mass of the apparently stationary star?

____

Luminosity, Brightness, and the Inverse Square Law

Name:________________________ **ID#:**__________________ **Date:**__________________

Luminosity is the total amount of energy per second given off by some object, such as a person, a lightbulb, or a star. Energy per second is usually measured in watts (for example, a 100-watt lightbulb uses twice as much energy per second as a 50-watt bulb of the same kind).

Brightness is the amount of energy per second **per unit area** received from some object. Two objects from which the same light energy per second per square centimeter reaches your eye will look equally bright to you. Brightness can be measured in watts per area, such as watts per square meter.

Imagine a lamp with two identical lightbulbs, one at the top (2 meters high) and one halfway up (1 meter high). Each lightbulb has an opaque shade that directs all the light from the lightbulb downwards as shown. Also shown are two circles on the floor around the lamp.

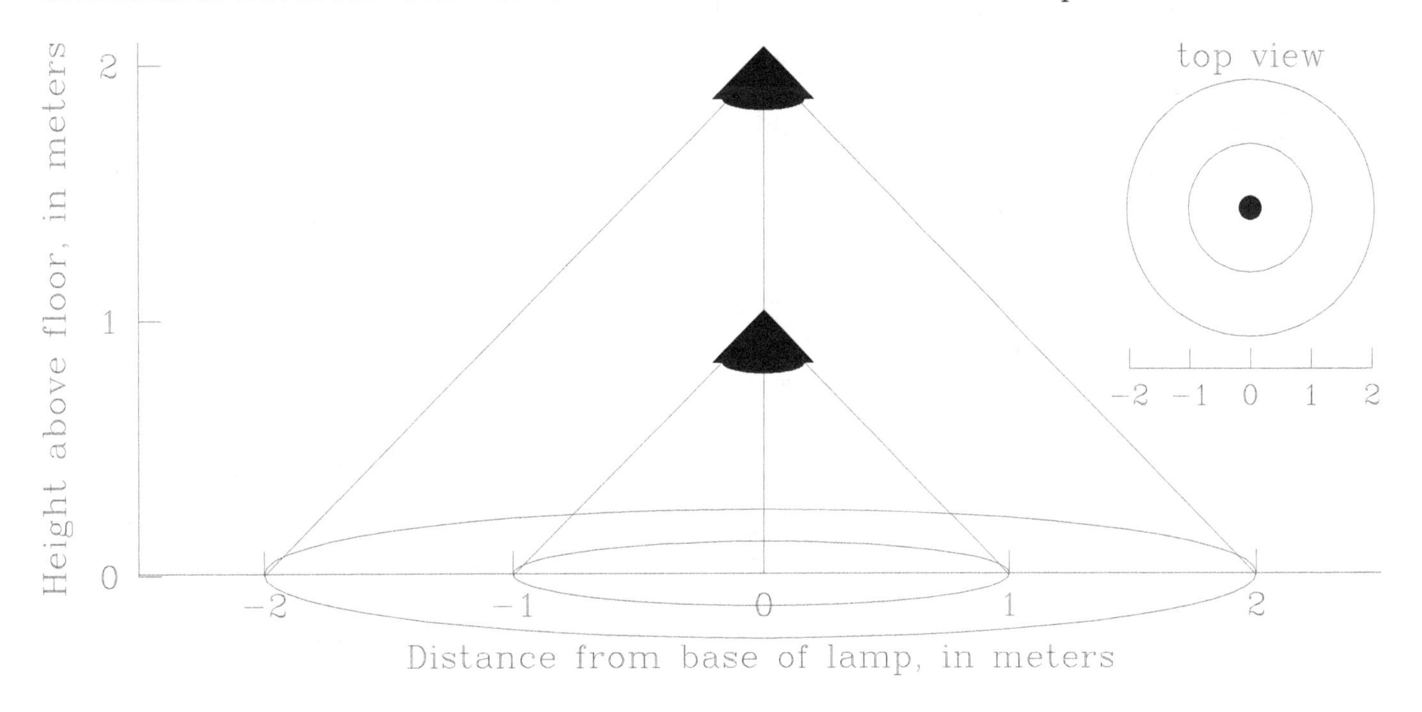

1. Suppose the lightbulb 1 meter above the floor is turned on. How far around the lamp in every direction does the light from that lightbulb hit the floor directly?

2. How many square meters of floor area are lit up by the lightbulb located 1 meter off the floor? Note that the area of a circle of distance r from center to edge is πr^2, where π is a number close to 3.14 and where $r^2 = r \times r$.

3. Suppose each lightbulb puts out 314 watts of light energy. On average, how many watts per square meter of light energy does the floor receive from this lightbulb? Assume a perfect mirror inside the shade, so that all the light from the bulb reaches the floor.

4. When the lightbulb 2 meters off the floor is turned on, how far around the lamp in every direction is the floor lit up? How many square meters of floor area are lit up in this case?

5. How many watts per square meter of light energy does that floor area receive, on average?

6. How many times less bright will the floor look (in a photograph, for example) as compared to when the lightbulb is 1 meter off the floor?

The Relationship Between a Star's Luminosity and Its Brightness

As the luminosity from a star travels out into space at the speed of light, it spreads out over an ever larger area. For example, a light-year is the distance travelled by light in a year. One year from now, the light emitted by the Sun in the past second will be spread out over the surface of a sphere measuring one light-year from center to surface (one light-year in radius). The area of the surface of a sphere of radius r is $4\pi r^2$.

Suppose Star One has a luminosity L and is located ten light-years away, and that Star Two also has a luminosity L but is located twenty light-years away, twice as far away as Star One. (This is the same situation we had with the two lamps on the previous page, just with two stars instead of two lightbulbs and with distances in light-years instead of meters.)

7a. Will Star Two look brighter or fainter than Star One?

7b. How much brighter or fainter will it appear? That is, if we measure a brightness of (say) 1 milliwatt per square kilometer for Star One, what will we measure for Star Two?

The brightness of a star falls off with increasing distance the same way the brightness of a lamp's light falls off with increasing height above the floor. Both brightnesses fall off with the inverse square of the distance: brightness B is proportional to luminosity L divided by distance^2, or, to be exact, $\boxed{B = L/4\pi r^2.}$ This relationship between B and L is an **inverse square law.**

Suppose Star One and Star Three appear equally bright in the night sky, but *parallax* measurements show that Star Three is three times as far away from us as Star One is.

8a. Which star is more luminous, the closer star (One) or the farther star (Three)?

8b. How do you know that star is more luminous than the other?

8c. How much more luminous than the other star is it?

9. **Bonus question:** How much larger is its parallax than the other star's parallax?

Elementary Particles and Forces Review

Name:________________________ **ID#:**_________________ **Date:**_________________

For the following questions, circle the correct answers:

1. The charge on the **proton** is (+, 0, −)
2. The charge on the **neutron** is (+, 0, −)
3. The charge on the **electron** is (+, 0, −)
4. According to Coulomb's law of electric force
   a. **Opposite** charges *(Attract | Repel)*

   b. **Like** charges *(Attract | Repel)*
5. In the hydrogen **atom** below, what force holds the electron to the proton? ________

e⁻ p⁺ Hydrogen Atom

6. In the helium **nucleus** drawn below what force holds the nucleus together? __________

p⁺ n⁰ n⁰ p⁺

7. In the helium nucleus above the **electric force** is trying to (hold the protons in the nucleus, push the proton out of the nucleus, push the neutrons out of the nucleus). (Circle correct answer.)
8. Can there ever be an *attractive* **nuclear force**: Between two protons? *(Yes | No)* Between two neutrons? *(Yes | No)* Between a proton and a neutron? *(Yes | No)*
9. What force holds you to the Earth? ______________________________
10. What force holds the moon to the Earth? __________________________
11. What force holds the Earth to the sun? ___________________________
12. What force holds you **up against** the force of gravity? ____________________

13. The answer to Question 12 tells you that the **electric force** is <u>(*Stronger* | *Weaker*)</u> than the gravitational force.

Fusion in the Sun: the Proton-Proton Chain

Name:________________________ **ID#:**________________ **Date:**________________

Introduction

When two nuclei in the center of a star collide at high velocity, they can fuse together to form a single larger nucleus.

If the mass of the new nucleus is less than the mass of the colliding nuclei, the mass difference is released into the star as energy. Fusion is the energy behind starlight.

The energy appears as some combination of **antielectrons** (which quickly collide with electrons and turn into photons), **photons**, **kinetic energy** (both of which provide energy to help keep the core of the Sun hot), or **neutrinos** (which escape the Sun immediately).

In the Sun, a new helium nucleus is formed by fusing together four hydrogen nuclei in several stages. In this activity you will work out the different pathways by which this can happen.

Figure 1: Four isotopes important to nuclear fusion in the Sun: hydrogen, heavy hydrogen (deuterium), light helium (helium-3), and helium. Each isotope is described by a letter, a small number giving the total number of protons or neutrons, and one or more plus signs. Each plus sign indicates one unit of positive electric charge.

Hydrogen	Deuterium	Helium−3	Helium
$^{1}H^{+}$	$^{2}D^{+}$	$^{3}He^{++}$	$^{4}He^{++}$
p	p n	p p n	p p n n
1 proton	1 proton 1 neutron	2 protons 1 neutron	2 protons 2 neutrons

Fusion Reactions

Two hydrogen nuclei can collide to produce a deuterium nucleus and by-products (an antielectron, which has the symbol e^{+}, and a neutrino), but that is not likely to occur every time such a collision happens. As a shorthand, we write:

$$^{1}H^{+} + {}^{1}H^{+} \longrightarrow {}^{2}D^{+} + e^{+} + \text{neutrino}$$

The number of positive charges is the same on each side of the arrow (net electric charge is conserved), and if you add up the small numbers on each side of the arrow, they are equal.

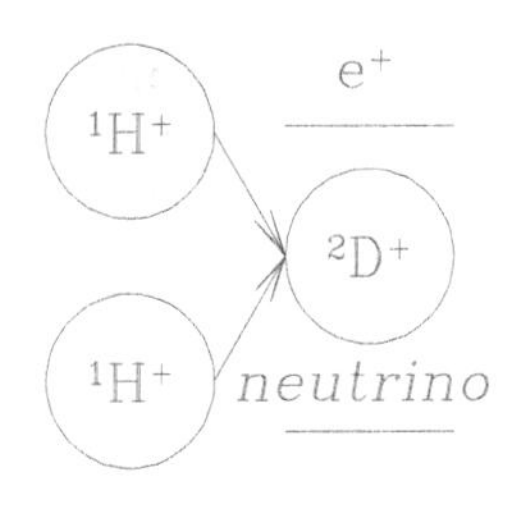

Figure 2: We can also draw this reaction as shown on the right. Each circle represents one nucleus. The arrows indicate that two $^{1}H^{+}$ nuclei collide and form one $^{2}D^{+}$ nucleus. By-products of the collision are written on the blank lines: in this case, an antielectron and a neutrino.

Fusion Pathways

To work out how four $^1H^+$ fuse into one $^4He^{++}$, use the following list of the important fusion reactions in the Sun and the approximate relative chance of each reaction occurring (the latter is the probability of the collision occurring times the probability of fusion happening in it):

$$^1H^+ + {}^1H^+ \longrightarrow {}^2D^+ + e^+ + \text{neutrino} \quad \textbf{(1 in 10 chance)}$$
$$^2D^+ + {}^1H^+ \longrightarrow {}^3He^{++} + \text{photon} \quad \textbf{(1 in 100 chance)}$$
$$^2D^+ + {}^2D^+ \longrightarrow {}^4He^{++} + \text{photon} \quad \textbf{(1 in 1,000,000 chance)}$$
$$^3He^{++} + {}^1H^+ \longrightarrow {}^4He^{++} + e^+ + \text{neutrino} \quad \textbf{(1 in 100,000 chance)}$$
$$^3He^{++} + {}^3He^{++} \longrightarrow {}^4He^{++} + {}^1H^+ + {}^1H^+ \quad \textbf{(1 in 2 chance)}$$

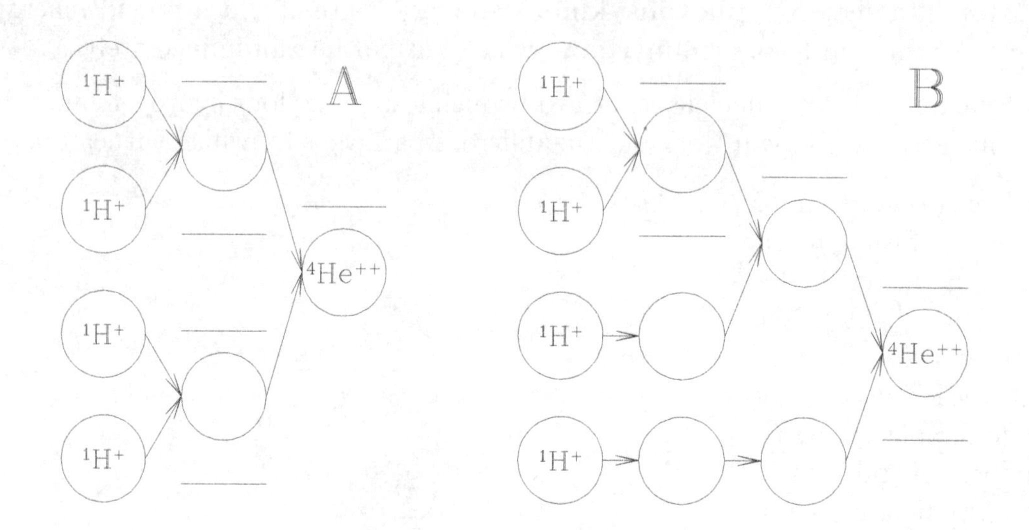

Figure 3: You can now fill in all the blank circles and lines in the three pathways to $^4He^{++}$ shown (A, B, and C).

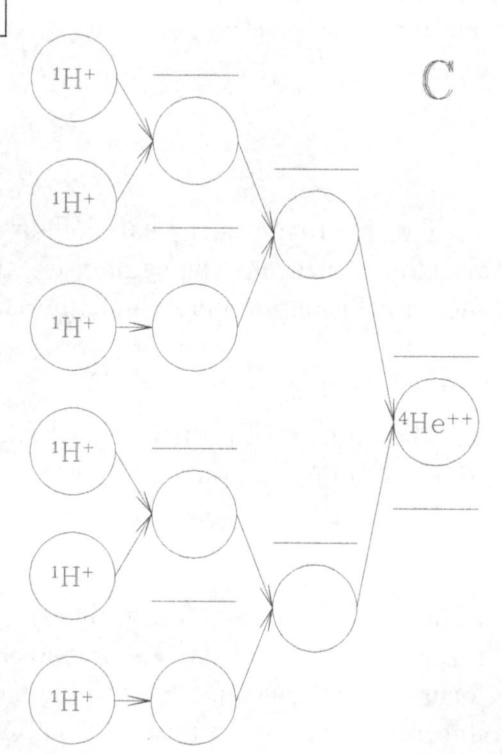

Fusion Pathway Rules:

I. The number in a circle must equal the sum of the numbers going into it.

II. The positive charge coming out of a collision must equal the positive charge going into it.

III. Only the five reactions at the top of the page are important here.

Questions to Answer:

1. How many neutrinos are produced along with every $^4He^{++}$ nucleus in pathway A ____ ? B ____ ? C ____ ?

2. Which of the three pathways will be the most common one? _______ How many times more $^4He^{++}$ nuclei will it produce than the others? ____________ and ____________

Spectra

Name:____________________ **ID#:**_______________ **Date:**_______________

1. Name the type of spectrum (the single form for spectra) an observer at A, B, or C would see if they were looking through a spectroscope.

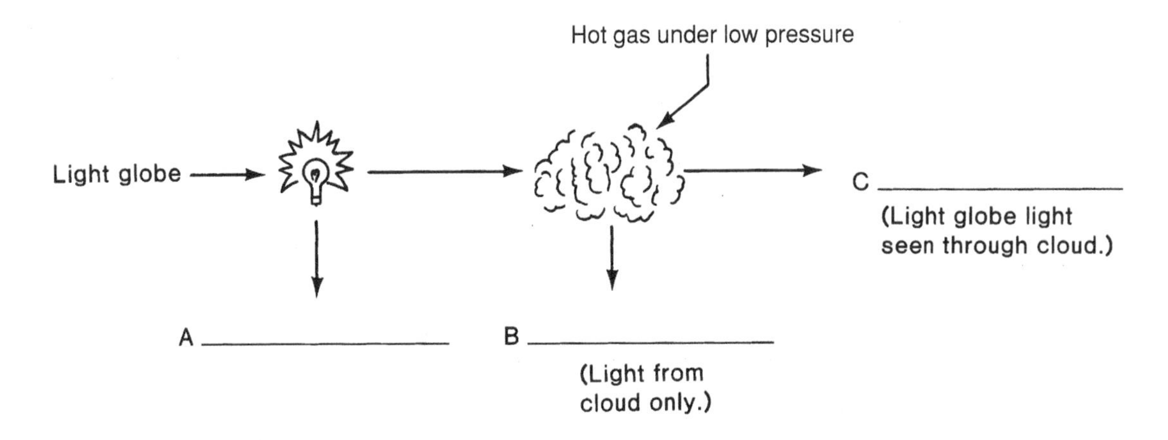

2. Below is a sketch of the cross section of a star.

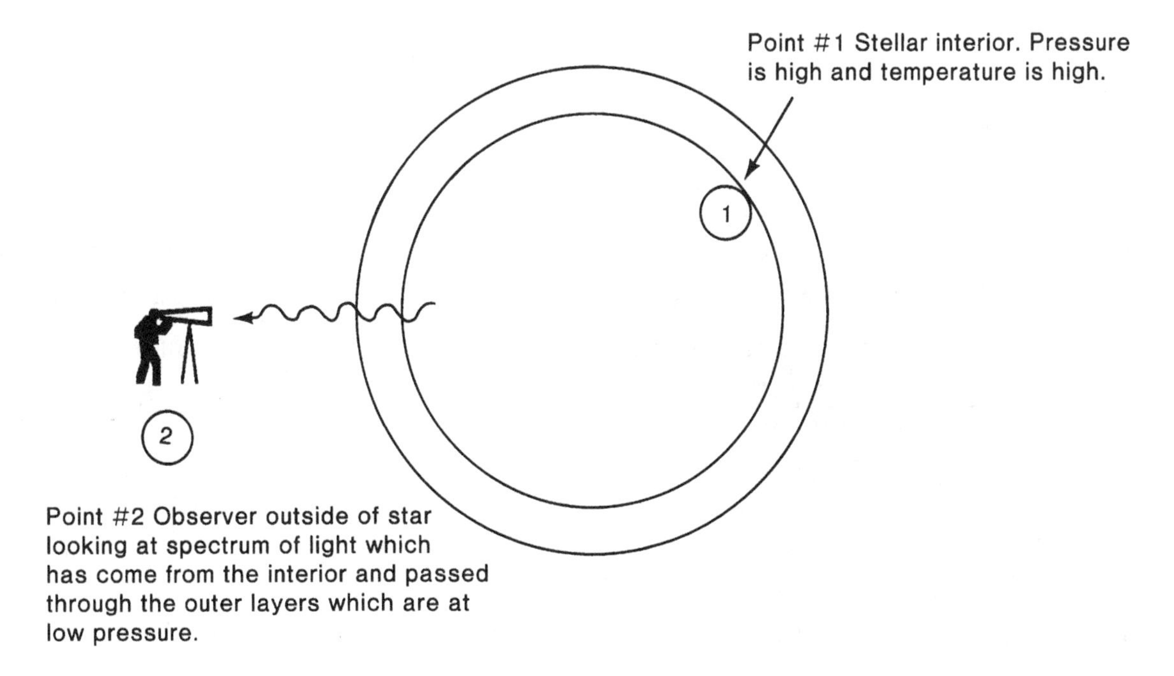

Figure 1: Cross Section of a Star

a. Is the pressure deep inside the star at Point #1 high or low? ____________________

b. What type of spectrum would an observer at Point #1 deep inside the star observe?

____

c. Is the pressure at the top of the star's atmosphere high or low? ______________

d. What type of spectrum would an observer at Point #2 outside the star looking in observe?

____

3. Which orbit must an electron in a hydrogen atom be in initially, in order for the hydrogen atom to absorb an Hα photon? ______________________________

4. Using an arrow, sketch the transition which an electron in the second orbit of a hydrogen atom would undergo when it absorbed an Hα photon.

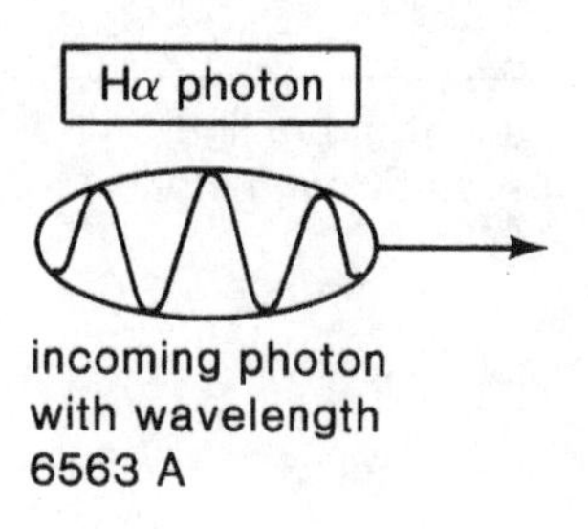

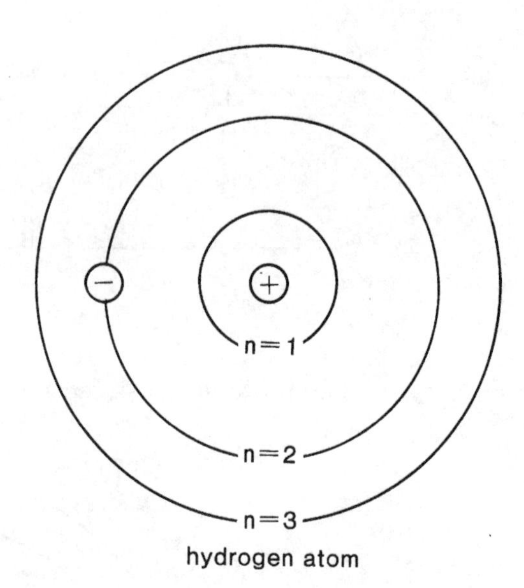

5. If the electron in the atom in Question 4 above jumped back to Bohr orbit #2, what type of photon would the atom emit? ______________________________

6. Because the pressure deep inside stars is always (*High* | *Low*), the interior has a(n) (*Absorption* | *Emission* | *Continuous*) spectrum.

a. The pressure at the top of a star's atmosphere is *always* (*High* | *Low*).

b. Therefore, when the spectrum from the interior of a star passes through the gas at ______________ pressure at the top of the star, a(n) ______________ spectrum will be produced.

c. Therefore all normal stars have ______________ spectra.

d. Therefore the sun must have an ______________ spectrum.

The Diameters And Luminosities of Stars

Objective

To learn about the vast size and luminosity differences between various stars located in the Solar System neighborhood. You will be able to draw models of stellar sizes and luminosities to scale to gain insight into their enormous range.

Introduction

Human scales for measuring distances and brightness – like Watts or kilometers – cannot properly capture the size and scale of many of the Universe's extreme objects. Hence, a scale model helps to envision the proportions of stellar sizes. Reading that the star Betelgeuse, located in the constellation Orion, has a diameter of about 5×10^8 or 500,000,000 km and emits about 16,000 times more energy into space than our Sun does seems meaningless given the staggering scale of these measurements. It is difficult for the mind to comprehend very large (as well as very small) numbers. If a pilot of a jet-liner tells you that you are now flying at an altitude of 37,000 feet, that statement may mean very little in terms of understanding the scope of the plane's height, as it is difficult to envision 37,000 feet. However, if we convert these 37,000 feet into miles, we come up with about 7 mile distance. And that means something to us; it is part of our everyday experience. So, to comprehend the vast scale of the Universe, we must make ourselves some models to the appropriate scale, comparing the diameters of stars to more recognizable objects, such as the diameter of the Earth.

Equations and Constants

Equation	Expression	Variables
Scale Factor Equation		*Scale:* the scale factor
	$Scale = \dfrac{Rule}{Real}$	*Real:* the actual size of a real-world object
		Ruler: the scaled down size of the real-world object
Radius	$R = \dfrac{D}{2}$	R: the radius of a spherical object D: the diameter of that object

Procedure

Worksheet #1, Model 1: This model compares planet Earth to some of the smaller stellar bodies. The 12,700 kilometer wide Earth is drawn as the small circle at the center of the page. Using a ruler, carefully measure the size of the scaled-down Earth. Using the Earth's real and scaled size as the starting point, calculate a scale factor. Using that scale factor, draw and label the following objects:

Barnard's Star

Sirius B - the smallest white dwarf

EG247 - the largest white dwarf

Moon's orbit around the Earth

Sun

Space these object out on the page (do not draw them as concentric circles). Use a compass to smoothly draw the circumference of those objects.

Worksheet #2, Model 2: This model compares the Sun – the largest object from Model 1 – to intermediate sized stars. The 1,500,000 kilometer wide Sun is represented by the small circle at the center of the page. As before, use a ruler to measure the diameter of the scale-model Sun. Using both the Sun's real and scaled size, calculate a new scale factor. Use that scale factor to draw the following properly proportioned objects:

Sirius A

Spica

Capella

Arcturus

Aldebaran

Mercury's orbit around the Sun

Worksheet #3, Model 3: this final model compares the size of our Solar System to some of the galaxy's largest stars. The small centered circle on this worksheet represents the diameter of the giant star Aldebaran (with a diameter 30× larger than the Sun). Once again, measure the diameter of Aldebaran with a ruler to obtain a ruler measurement. Using the 30 $D_{\odot}$ measurement for the real diameter of Aldebaran, calculate a new scale factor. Use that scale factor to draw the following properly proportioned objects:

Earth's orbit around the Sun

Mars's orbit around the Sun

Jupiter's orbit around the Sun

Rigel

Antares

Betelgeuse

For the orbits of the planets, draw those as concentric circles centered at Aldebaran to demonstrate the size of planetary orbits in relation to very large stars. Draw Rigel, Antares, and Betelgeuse on their own.

The next portion of the lab – Worksheets #4 through #7 – deals with the light output of these stars. The visible light emitted by a star is known as the luminosity (although, strictly speaking, this term also encompasses light from the entire electromagnetic spectrum), and is normally measured in units of solar luminosities (with the Sun's light output set to 1 $L_{\odot}$). Much like the physical sizes of the stars vary so greatly, a single on-paper model cannot reasonably represent the luminosity of both the lowest and highest luminosity stars.

Use Worksheet #4 to represent the scaled light output of each of the given stars on three different scales: a scale for low-luminosity stars, where the Sun's light output dominates, a scale for intermediate-luminosity stars, where the Sun is a small fraction of the light output, and a scale for high-luminosity stars, where the Sun's light output in a very small fraction only. The first model compares the luminosity of the Sun to the light output of the following small, low mass, low intensity stars close to the Solar System:

Barnard's Star, Kapteyn's Star, 61 Cygni, ε Eridani, and τ Ceti

The Sun's luminosity serves as the baseline for comparison and is scaled such that 1 $L_{\odot} = 100$ square centimeters (10 cm across by 10 cm high; this is also equivalent to 10,000 of the small boxes, each of which measures a square millimeter). Worksheet #6 contains a 10 cm by 20 cm graph (in total, there are 200 cm^2 surface area composed of 20,000 small 1-millimeter by 1-millimeter boxes). The Sun's luminosity is represented by the box labeled "Sun," taking up a full half of the graph. Using the remaining boxes, mark off areas representing the light output of the other 5 stars.

Worksheet #7 compares the light output of the Sun to the moderately bright stars Capella and Sirius A. As on Worksheet #6, the Sun's luminosity is already drawn. However, rather than taking up half of the graph, the Sun's luminosity only occupies a 1-cm by 1-cm square (10-mm by 10-mm). Mark off rectangular areas to represent the light output of Sirius A and Capella.

Finally, Worksheet #8 compares the light output of some of the brightest stars. Now, the Sun's luminosity has been reduced to a single, tiny, 1-mm by 1-mm box. Each one square centimeter box represents 100 $L_{\odot}$. Using this new and final scale, mark off rectangular blocks to represent the light output of the following stars:

Sirius A, Capella, Antares, Betelgeuse, and Rigel

Datasheet #1

The following table conveys some sizes which will be needed for scaling the size of major stars in Worksheets #1, #2, and #3.

Object	Measurement	
Solar Diameter	1,500,000 km	
Earth Diameter	12,700 km	
Diameter of the Moon's Orbit	770,000 km	
Diameter of Mercury's Orbit	120,000,000 km	0.4 AU
Diameter of Earth's Orbit	300,000,000 km	1.0 AU
Diameter of Mars's Orbit	450,000,000 km	1.5 AU
Diameter of Jupiter's Orbit	1,560,000,000 km	5.2 AU

Star Name	Diameter in Solar Diameters
Largest/Smallest White Dwarf	0.02/0.0067
Barnard's Star	0.1
τ Ceti	0.8
Sirius A	1.8
Spica	7.5
Capella	12
Arcturus	20
Aldebaran	30
Antares	600
Betelgeuse	800

Worksheet # 1

NAME ______________________________ ID# ______________________________

DATE ______________________________ LAB SECTION# ____________________

Model 1: Sizes of Small Stars

Worksheet # 2

NAME ______________________________ ID# ______________________________

DATE ______________________________ LAB SECTION# ____________________

Model 2: Sizes of the Sun and Intermediate Stars

Worksheet # 3

NAME ______________________________ **ID#** ______________________________

DATE ______________________________ **LAB SECTION#** ____________________

Model 3: Sizes of Large Stars

Worksheet # 4

NAME ______________________ ID# ______________________

DATE ______________________ LAB SECTION# ______________________

		Scale Factors				
		Model 5		Model 6	Model 7	
Star Name	**$L/L_{\odot}$**	**cm^2**	**mm^2**	**cm^2**	**cm^2**	**mm^2**
Sun	1	100	10,000	1	1/100	1
Barnard's Star	0.0005					
Kapteyn's Star	0.004					
61 Cygni	0.08					
ε Eridani	0.3					
τ Ceti	0.5					
Sirius A	24					
Capella	150					
Antares	5,500					
Betelgeuse	13,800					
Rigel	60,000					

1. How many small, millimeter sized boxes are located on the graph on Worksheet #7?

2. How many small, millimeter sized boxes are required to show the light output of the supergiant star Rigel?

3. What percentage of Rigel's total light output will fit on the page on Worksheet #7?

Worksheet # 5

NAME ______________________ ID# ______________________

DATE ______________________ LAB SECTION# ______________________

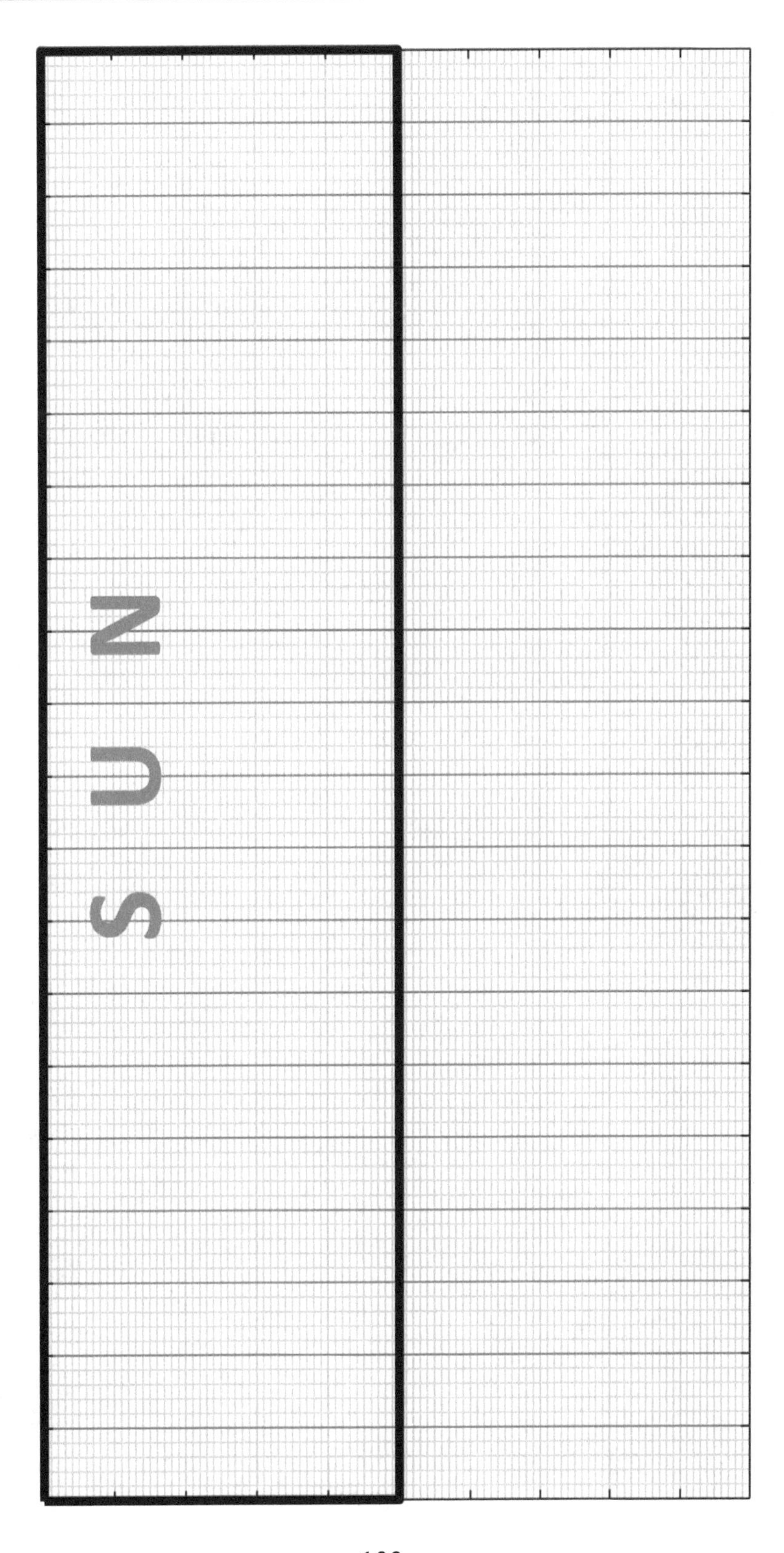

Worksheet # 6

NAME ______________________________ ID# ________________________________

DATE ______________________________ LAB SECTION# __________________

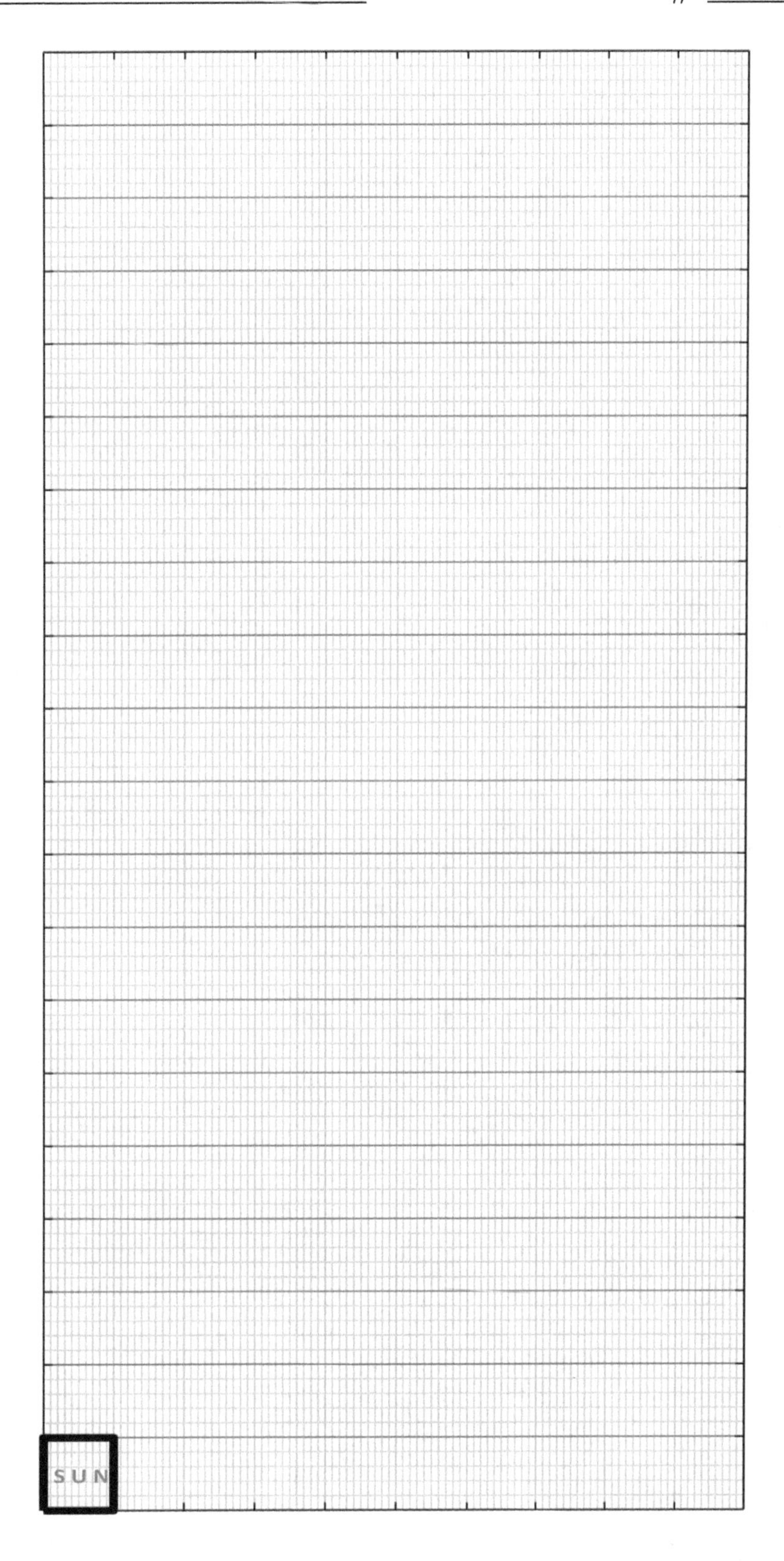

Worksheet # 7

NAME ______________________ ID# ______________________

DATE ______________________ LAB SECTION# ______________

Stellar Classification

Name:________________________ **ID#:**________________ **Date:**________________

"In questions of science, the authority of a thousand is not worth the humble reasoning of a single individual." *Galileo*

Part I: Introduction

1. H-R Diagram Review

> **Note:** All main sequence stars have almost the same chemical composition. Hydrogen is the most abundant and helium is the second most abundant element in their atmospheres.

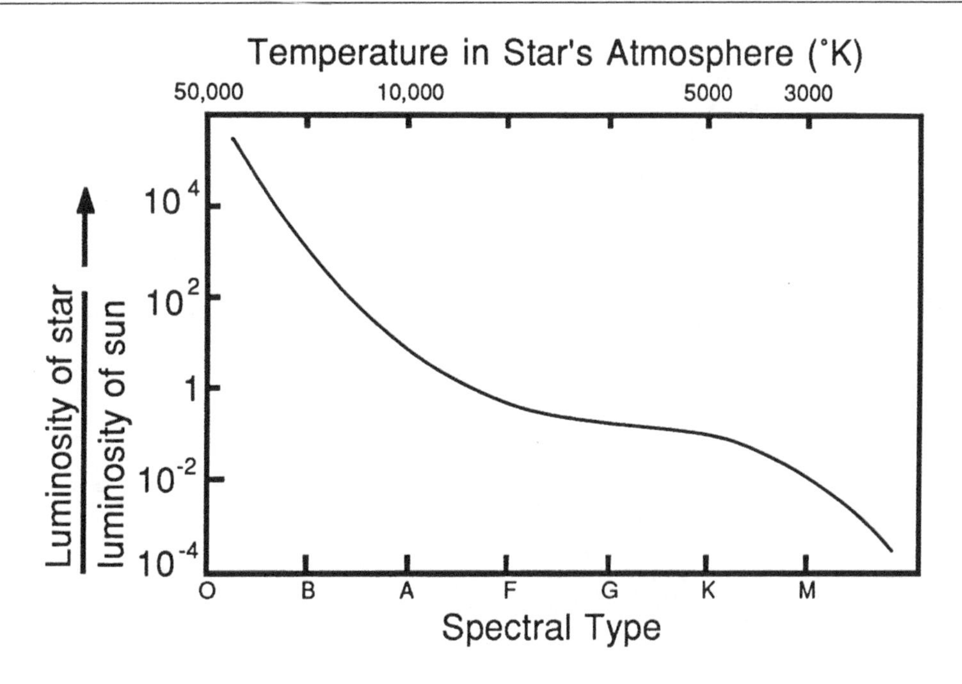

Figure 1: The Hertzsprung-Russell Diagram

a. On the H-R diagram above label where red giants are found.

b. On the H-R diagram above label where white dwarfs are found.

c. Which main sequence spectral type is hottest? ____________________

d. Which main sequence spectral type is coolest? ____________________

e. Antares is a red supergiant with an atmospheric temperature of 3100° K. What is Antares' spectral type? ____________________

f. Sirius B is a white dwarf with an atmospheric temperature of 10,000° K. What is Sirius B's spectral type? ____________________

The Hertzsprung-Russell (H-R) Diagram

Name:__________________________ **ID#:**__________________ **Date:**__________________

Every star has a **surface temperature** (T) and a **luminosity** (L) (energy output per second). A star's temperature is easy to measure. Measuring a star's luminosity requires knowing its distance, which is not easy to measure accurately except for relatively nearby stars.

If a star's luminosity and temperature have been measured, we can plot it on a **Hertzsprung-Russell diagram**, or H-R diagram for short. The H-R diagram is a plot of luminosity in the vertical direction versus temperature (or color, or spectral type) in the horizontal direction. **Every star can be plotted on the H-R diagram,** just like every person can be plotted on a chart of height vs. weight.

As a star ages, its temperature and color can change. Therefore, **a star appears in different places on the H-R diagram at different stages of its life,** just like people would appear in different places on a height vs. weight chart at different stages of their lives.

Make Your Own H-R Diagram

1. You have measured the luminosities and colors of five stars in Star Cluster Yamato. Using a standard color scheme to convert from color to temperature, we observe that stars that are seen to have one of seven color shades are known to have one of the seven temperatures shown in the table below. Each star also has a unique luminosity relative to that of our Sun's luminosity.

Use the following seven color shades to fill in the table: blue-white, orange, red, red-orange, white, yellow, and yellow-white. Recall that blue-white stars are the hottest, with temperatures of 14000 K. **Stars of lower temperatures have different colors in the same way that wood burning in a fire has different colors as the fire cools down.**

Color:				yellow			blue-white
Temperature in K:	3300	4000	5000	5800	7000	10000	14000
Luminosity vs Sun's:	1/1000	1/65	1/4	1	4	60	500

2. For each of the five stars from Star Cluster Yamato in the table below, fill in the temperature and luminosity using the completed table from Question 1.

Star	Color	Temperature T, in K	Luminosity L relative to the Sun's	(**Optional**) Radius R relative to the Sun's
A	yellow			
B	yellow-white			
C	red-orange			
D	red			
E	orange			
Sun	yellow	5800	1	1

3. (**Optional**) Given the temperature and luminosity of each star in the previous table, work out the radius R of each of the stars. Use the relationship $L/L_{Sun} = (R/R_{Sun})^2(T/T_{Sun})^4$.

4. Use the completed table in Question 2 to plot the stars on the H-R diagram in Figure 1. Then read and answer the questions on the last page of this activity.

The three lines on the H-R diagram in Figure 1 below show where **main-sequence stars** of three different ages will be found. **The parts of the main sequence where stars from a particular star cluster are found depends on the cluster's age,** just like the position of a group of people born the same year on a height vs. weight chart depends on their age.

In detail:

- The **short-dashed line** shows where 50 million-year-old main-sequence stars would be found.
- The **solid line** shows where 3 billion-year-old main-sequence stars would be found.
- The **long-dashed line** shows where 20 billion-year-old main-sequence stars would be found.

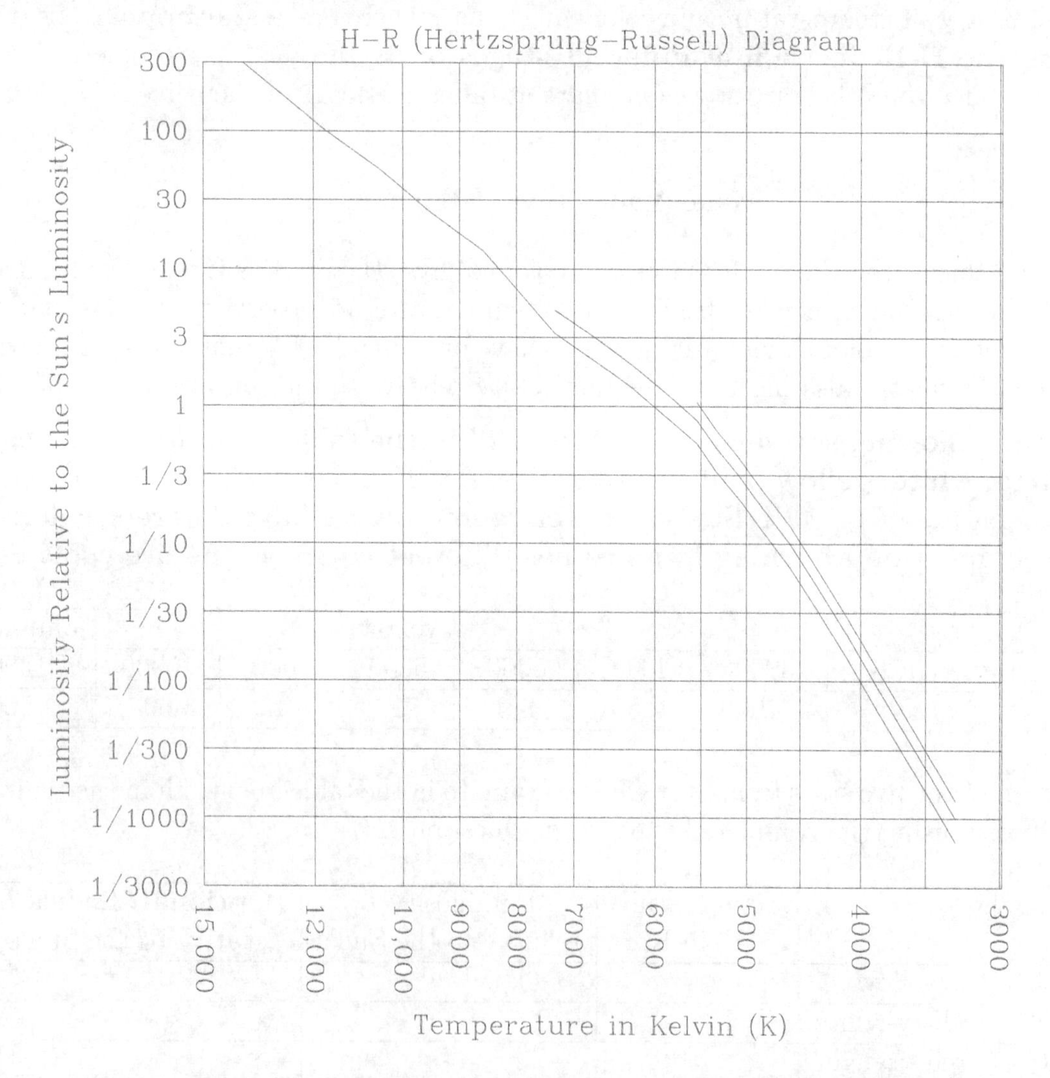

Figure 1: A Hertzsprung-Russell diagram. The luminosity-temperature relationships shown as the three lines are approximations to the real relationships.

Name:________________________ **ID#:**________________ **Date:**________________

5. Are the stars you studied in Star Cluster Yamato main-sequence stars? *(Yes | No)*

6. Which of the following ages is the best match to the age of Star Cluster Yamato?

a) 50 million years old

b) 3 billion years old

c) 20 billion years old

7. Star Cluster Yamato contains some **giant** stars. Indicate the region in Figure 1 where some giant stars could be found.

8. Star Cluster Yamato also contains **white dwarfs**. Indicate the region in Figure 1 where some white dwarfs could be found.

9. As a white dwarf ages, its position on the H-R diagram changes. Indicate on Figure 1 the direction in which a white dwarf moves over time.

10. (**Optional**) Draw on Figure 1 a straight line that connects all stars with a radius equal to the Sun's radius. Do the same for all stars with a radius 10 times the Sun's radius and for all stars with a radius $\frac{1}{10}$ times the Sun's radius. (Hint: See Question 3.)

Stellar Evolution

Name:______________________ **ID#:**________________ **Date:**________________

"I believe a leaf of grass is no less than the journeywork of the stars."
Walt Whitman, *Leaves of Grass*, 1855

Part I: The Role of Forces, Temperature, and Energy in Star Formation

I. **Gravity** pulls gas and dust clouds together to form stars and liberates gravitational potential energy as the cloud collapses.

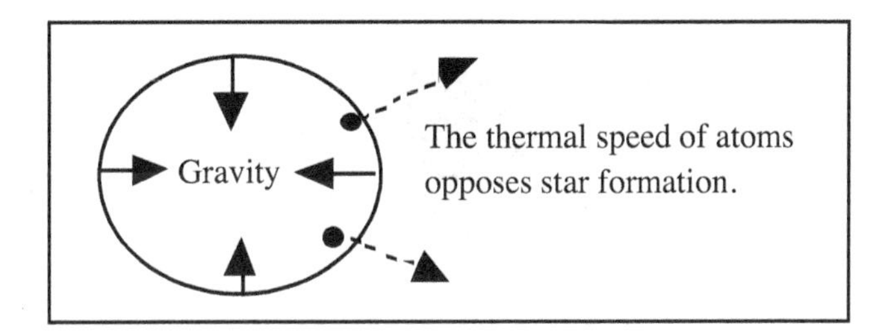

1. As the cloud collapses *PE* (**Potential Energy**) → ________ → ________ and T_{core} (⇓ ⇑).

II. **Temperature** controls the **thermal speed** of atoms and molecules and **opposes** star formation.

$$V_{atoms} = \sqrt{\frac{3kT}{m_{atom}}}$$

where V_{atom} is the speed of the atom, k is the Boltzmann constant, T is the temperature and m_{atom} is the mass of the atom.

2. According to this equation, as the temperature of the cloud ⇑, the speed of atoms (⇓ ⇑).

3. This means that star formation is more likely in clouds that are (*Hot* | *Cold*).

III. **Pressure** slows down the collapse of the cloud and ultimately after pressure counterbalances gravity the cloud **stops** collapsing.

Pressure is the force/area. In everyday units the atmospheric pressure on your body is 14.7 lbs/in^2. Your weight causes pressure on the bottom of your feet. A 150-pound person with each foot having an area of 40 in^2 would feel a pressure of 1.9 lbs/in^2 on the bottom of **each** foot.

4. If this 150-pound person stood on **one** foot, what would be the pressure on the bottom of that foot? ____

The **thermal speed** of atoms and their number density control the pressure, P, within the cloud. **Pressure** is given by the equation

$$\mathbf{P = nkT,}$$

where $\boldsymbol{n}$ is the number density of atoms (atoms/km^3), k is Boltzmann's constant and $\boldsymbol{T}$ is the temperature.

5. As the cloud collapses n ⇑ and T ⇑. The equation above tells us that P must ______________ until eventually the pressure **stops** the collapse of the cloud.

First Two Stages of Stellar Evolution

Stage I Protostars

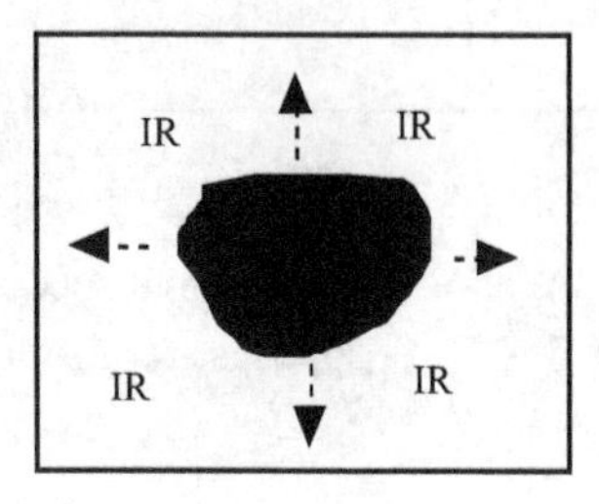

In the Protostar stage the cloud is so cold that it is dark and only emits IR **radiation**.

1. What force makes the cloud collapse? ______________________
2. What is the energy source for Protostars? ___________________
   a. As the cloud collapses T_{core} (⇓ ⇑).

   Because the temperature and pressure in interstellar space is very low, initially the pressure in the cloud is **essentially zero** and the cloud is in a state of **free-fall**.
3. As the cloud collapses the pressure (⇓ ⇑) and the rate of collapse (*Increases* | *Decreases*) until eventually the force of gravity is counterbalanced by the ______________ and the **cloud stops collapsing**.

Stage II T Tauri

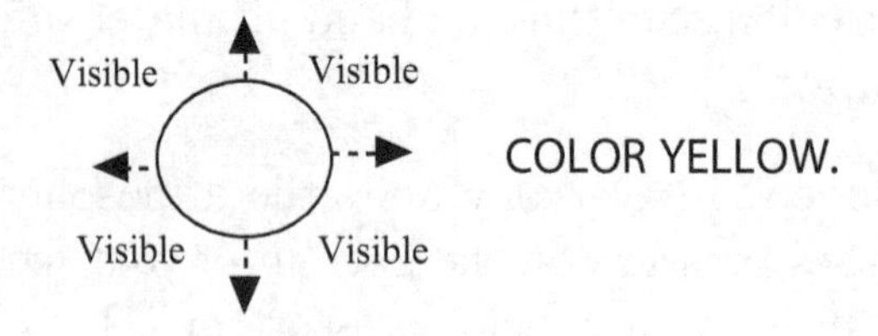

In the T Tauri stage the cloud is hot enough to emit visible radiation.

1. What is the energy source for T Tauri stars? ______________________________
   a. As the T Tauri star collapses T_{core} (⇓ ⇑).
2. What is the primary reason that fusion does **not** occur in Protostars and T Tauri stars?

   ____

Part II: The Interplay of the Electric Force, Nuclear Force, and Temperature Determines When Fusion Begins

I. Electric force is governed by Coulomb's law.

$$F = \frac{C\,Q_1 Q_2}{d^2}$$

where F is the electric force between two charges Q_1 and Q_2 separated by the distance d. C is a constant.

Coulomb's law tells us that

1. **like charges (two protons)** (*Attract* | *Repel*) and **opposite charges (an electron and a proton)** (*Attract* | *Repel*).

Add an arrow on each proton on the left showing the **direction** of the force between the two protons.

2. as two protons get **closer together (d smaller)** the repulsive force between them (*Increases* | *Decreases*).
3. as the **amount** of **like** charge increases (Q_1 and Q_2 larger) the repulsive force (*Increases* | *Decreases*).

II. **Nuclear force** acts between protons and neutrons and depends on the distance between them.

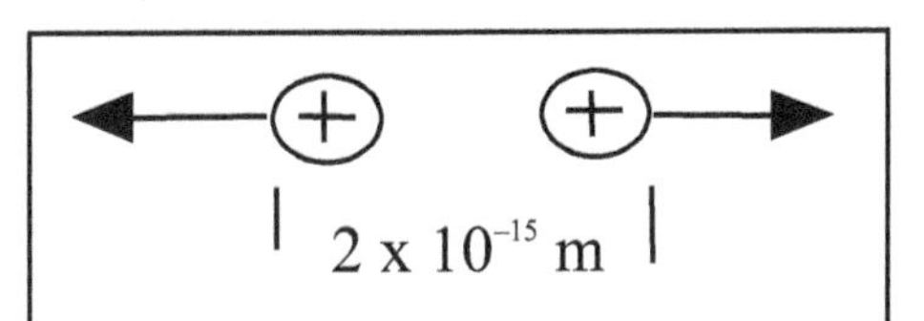

In the fusion reaction protons are pushed together. When the distance between two protons becomes **less** than 2×10^{-15} m they **stick together**.

4. What force makes it difficult to push two protons together? ________________
5. 2×10^{-15} m is a very **small** distance. **Coulomb's law** tells us that it will require a very (*Large* | *Small*) impact speed to push two protons this close together.

III. **Temperature** controls the speed of protons in stars.

$$V_{protons} = \sqrt{\frac{3kT}{m_{protons}}}$$

where V_{proton} is the speed of a proton, k is Boltzmann's constant, T is the temperature and m_{proton} is the mass of a proton.

6. According to this equation, as T ($\Uparrow$) $V_{protons}$ ($\Downarrow$ $\Uparrow$) and protons strike each other (*Harder* | *Softer*) and when they collide they will get (*Closer together* | *Farther apart*).

7. At $T = 10$ **million K** when two protons collide they are **less** than 2×10^{-15} m apart. What will the two protons do? (*Stick together* | *Bounce apart*) ________

Stage III Main Sequence: a star is born. **Our sun** is in this stage.

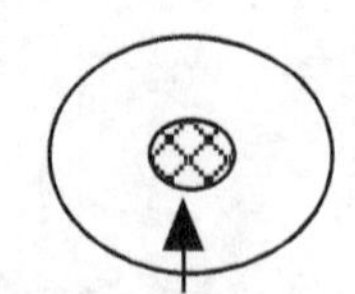

COLOR YELLOW.

Fusion is occurring in core.

8. **Fusion 4 H → He + energy** is occurring in main sequence stars, because the temperature in the core **is more than** ________________________.

9. What element is being manufactured in the core of main sequence stars? ________________ What is the fuel for the fusion reaction? ________________

10. After the hydrogen in the core of the main sequence star is **exhausted**, the core contracts. As the core contracts the radius, R, of the core ($\Downarrow \Uparrow$) and PE → ____ → ____ and T_{core} ($\Downarrow \Uparrow$).

    When the temperature in the core reaches **100 million K** the star moves to Stage IV the Red Giant Stage.

Part III: Why It Requires a Higher Temperature (100 Million K) for the Triple Alpha Process than the Fusion Process (10 Million K)

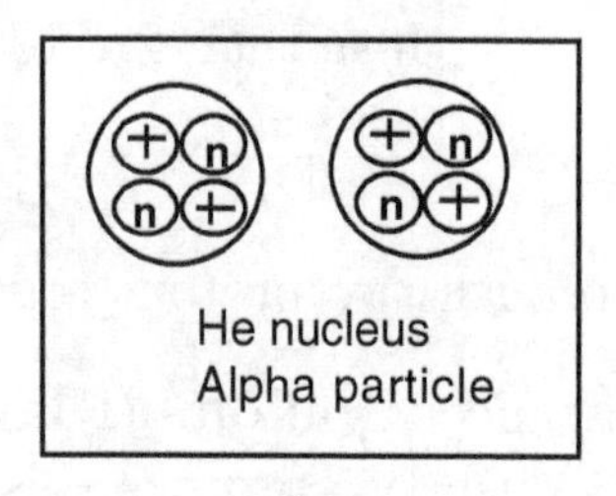

11. How many protons are in each He nucleus? ________________

12. Because there is **more charge** in the helium nucleus than in the hydrogen nucleus, there is (*More* | *Less*) **repulsion** between the two helium nuclei.

13. **More repulsion** means it will take (*Higher* | *Lower*) **speeds** to push the He nuclei together and make the triple alpha process begin. This means that it will require a (*Higher* | *Lower*) **temperature** to attain these speeds.

14. Now explain why the triple alpha process **3 He → C + energy** requires a temperature 100 million K while fusion **4 H→ he + energy** only requires a temperature of 10 million K.

Stage IV Red Giant Stage

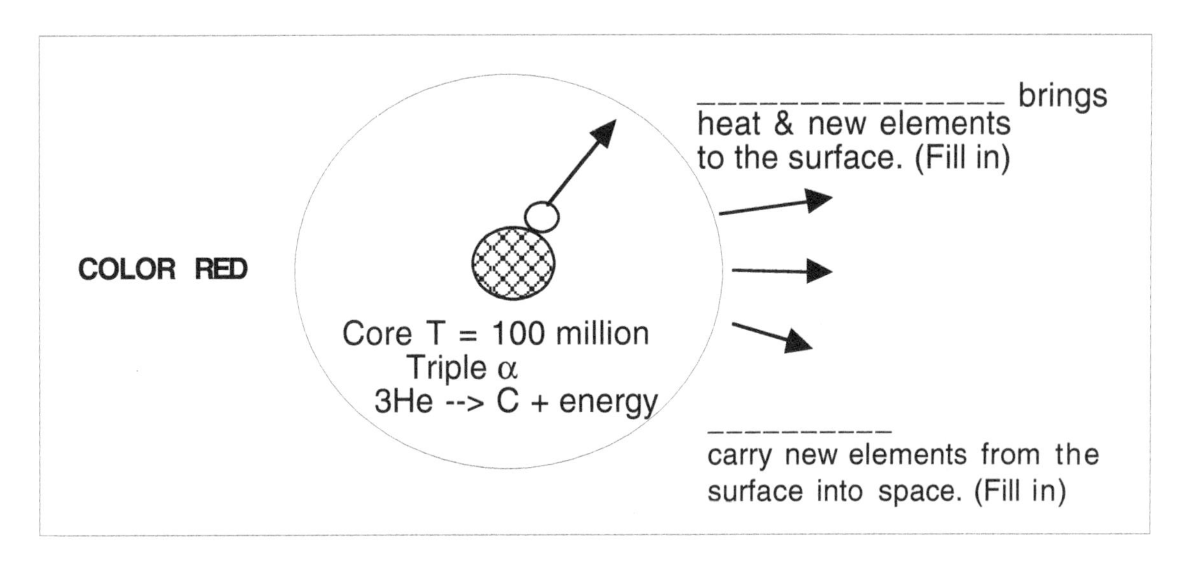

15. What happens to a gas when it is heated? ____________________ Explain why our sun will expand out to the **orbit of Venus** when it becomes a red giant.

    a. What will happen to Mercury and Venus when our sun becomes a red giant? ____________________ What will happen to the temperature of the Earth? ____________________ What do you think will happen to the Earth's oceans by that time? ____________________

16. The **velocity of escape** from a star is given by $V_{escape} = \sqrt{\frac{2GM_{star}}{R_{star}}}$, where G is the universal gravitational constant, M_{star} is the mass of the star and R_{star} is the radius of the star. Explain why red giants have **stellar winds** that are much stronger than main sequence stars.

17. Red giants make all of the elements heavier than He up to and including Fe (iron). Explain why Fe is the **heaviest element** that is made in the core of red giants.

Special Relativity: Time Dilation and Length Contraction

Name:________________________ **ID#:**________________ **Date:**________________

The key concept of **Einstein's Special Theory of Relativity** (or **special relativity** for short) is that **light travels at the same speed for everyone**. Accepting that result led Einstein to new insights about our weird universe.

We can study these strange but true effects of special relativity by doing thought experiments involving rocket trains that travel at speeds more than 10% of the speed of light. If we really did routinely travel at such speeds, we would notice the effects of special relativity every day; they would seem normal, not strange.

Vertical Laser Clocks and Time Dilation

Our first thought experiment will show that **time is relative**. All we need to know is that time = distance divided by speed and that the speed of light (c) is about 300 million meters per second, or 1 foot per nanosecond (ns). A nanosecond is one billionth (10^{-9}) of a second.

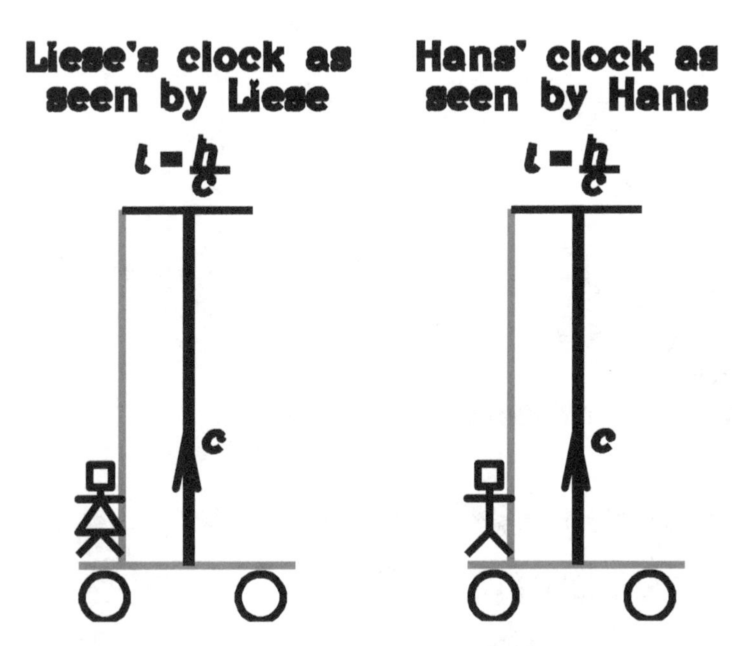

Figure 1: We have prepared two identical train cars on two separate rocket trains. Liese is in the train on the left and Hans is in the train on the right. Neither train is moving in this figure. Each train car is equipped with a laser clock: a laser mounted on the floor of the car facing a mirror mounted a height (h) above the laser. Both Liese and Hans see a pulse of laser light launched upwards at speed c (arrow and dashed line), reflect off the mirror after time $t = \frac{h}{c}$ to make one *tick* of this clock, and return (following the dotted line) to the floor as the next *tick* of this clock. As soon as the laser pulse returns to the floor, a new one is sent out, making a very accurate clock.

Now consider what happens when Hans' rocket train is moving to the right at speed v relative to Liese's rocket train. Figure 2 on the next page shows Hans' train car at three different times: when a laser pulse is launched, when it bounces off the mirror, and when it returns to the floor.

Liese's clock runs the same as in Figure 1: it makes a tick every $t = \frac{h}{c}$ seconds.

But now Liese sees Hans' rocket car and everything in it moving to the right with speed v. Because of that, Liese sees the light in Hans' laser clock travel farther to reach its mirror. By the time Hans' light bounces off its mirror, that light has traveled a distance d (dashed line) which is greater than h. **Because light always travels at the same speed for everyone, Liese sees a longer time between ticks on Hans' clock**: $t' = \frac{d}{c}$.

In other words, time is relative, and **moving clocks run slow**. That is, from your point of view, **clocks that are moving relative to you run slower than clocks that are not moving relative to you.** This effect is known as time dilation.

We can determine how much slower Liese sees Hans' clock running by computing the distance d. In the time t' that it takes Hans' light to travel from floor to mirror, Hans' train car has moved a distance vt' (Figure 2, bottom right). The distances d, h, and vt' form a right triangle, for which the Pythagorean theorem tells us that $d^2 = (vt')^2 + h^2$. We also know that $d = ct'$. Substituting in for d and solving for t', we find that $\boxed{t' = t/\sqrt{1-(v/c)^2}}$.

1. If Hans' speed is $v = 0$, guess how would t' compare to t: ____________________
Plug $v = 0$ into the equation for t' above to confirm your guess (show your work):

2. If Hans' speed v is very close to the speed of light c, then $1-(v/c)^2$ is a very small number. Will t' be much larger or much smaller than t if v is very close to c? ________________ Therefore, the faster Hans' rocket train moves, the ____________ Liese sees Hans' clock tick.

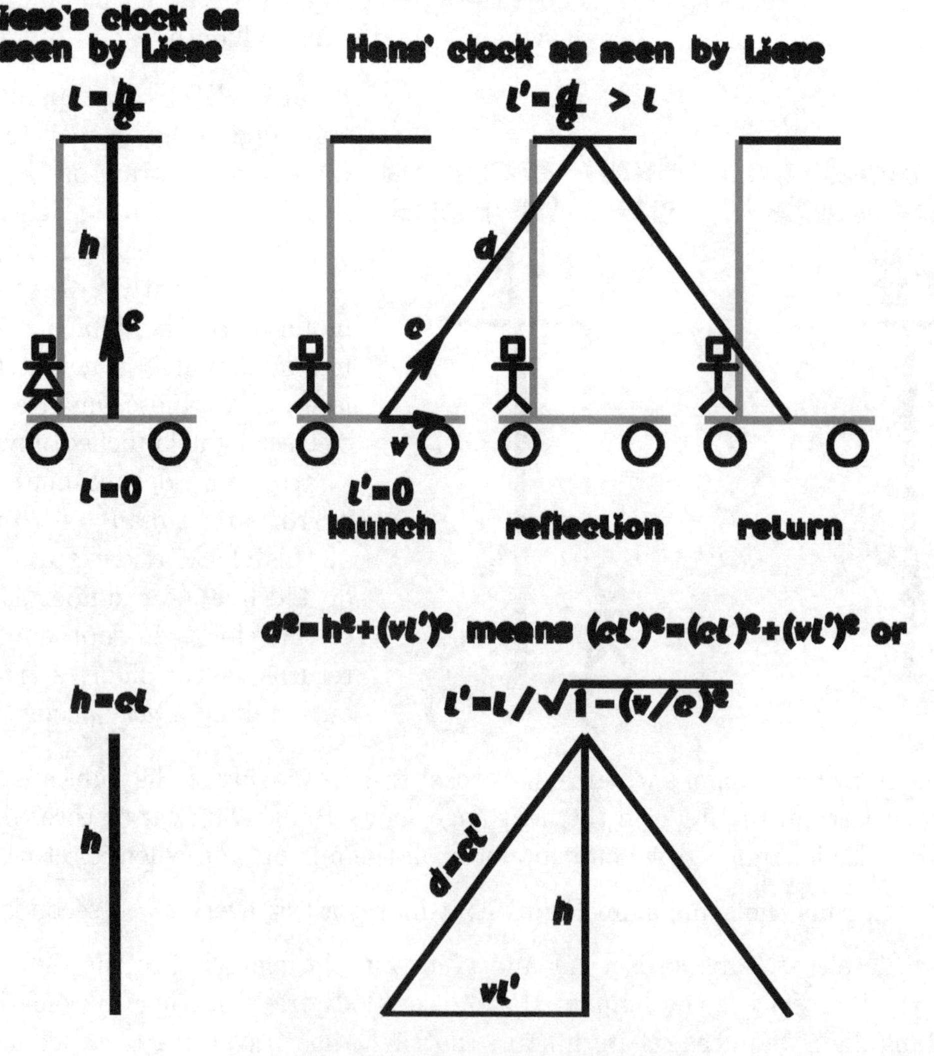

Figure 2: Hans' clock moving relative to Liese.

Now consider the same situation from Hans' point of view (Figure 3). From Hans' point of view, he is not moving and Liese is moving to the left with speed v. Therefore, **the exact same analysis we did for the situation in Figure 2 applies in reverse.** Now Hans sees his clock ticking the same as it did in Figure 1 and he sees Liese's clock run slowly. This is another example of how **clocks moving relative to you run more slowly from your point of view than clocks not moving relative to you.**

Keep in mind that **Figures 2 and 3 show the exact same situation, just from two different perspectives. Therefore, Liese sees Hans' clock run slow while Hans sees Liese's clock run slow.** And note that this result applies to all clocks, not just laser clocks.

3. Whose clock is keeping the correct time: Liese's, Hans', neither, or both? Justify your answer.

___

___

___

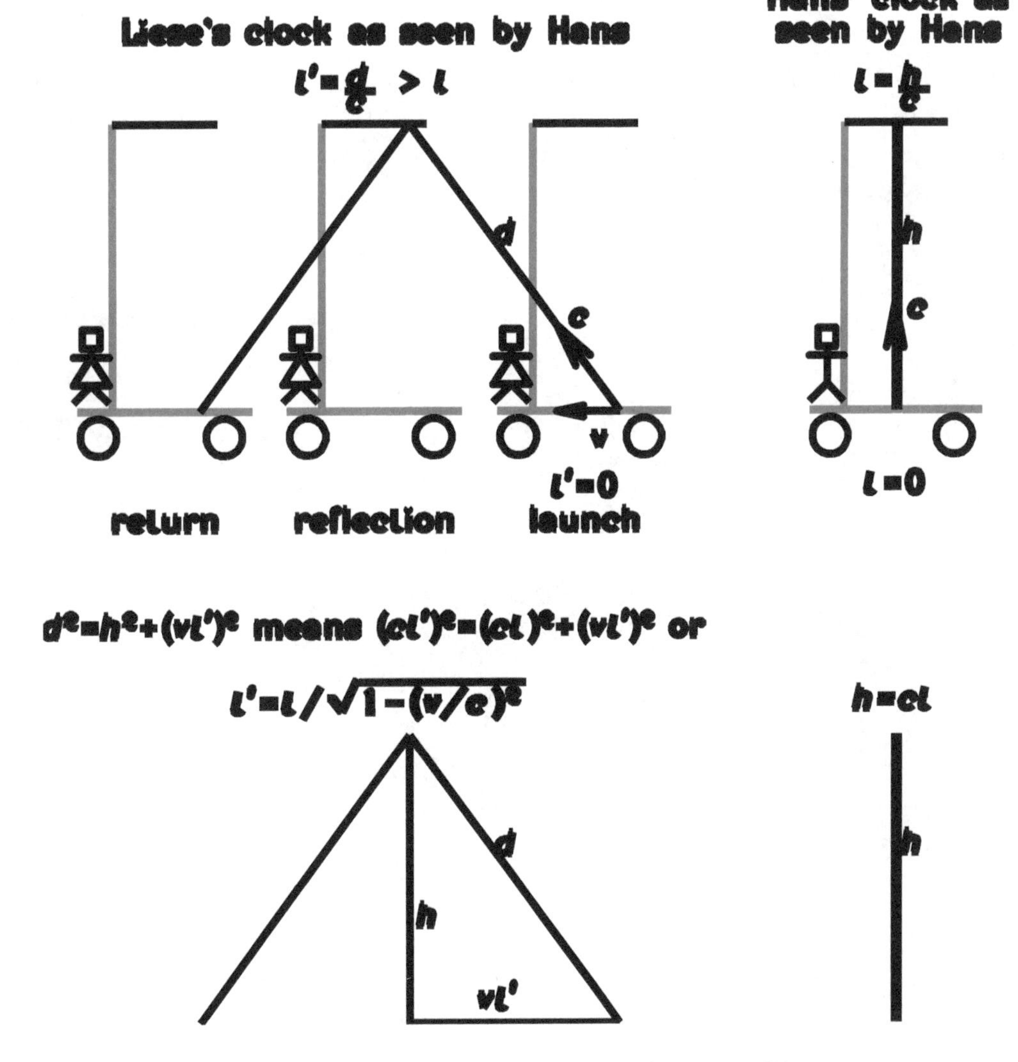

Figure 3: Liese's clock moving relative to Hans.

Horizontal Laser Clocks and Length Contraction

Our next thought experiment involves laser clocks placed horizontally. They are otherwise the same as before: a pulse of laser light is fired at a mirror, reflects off the mirror, and returns.

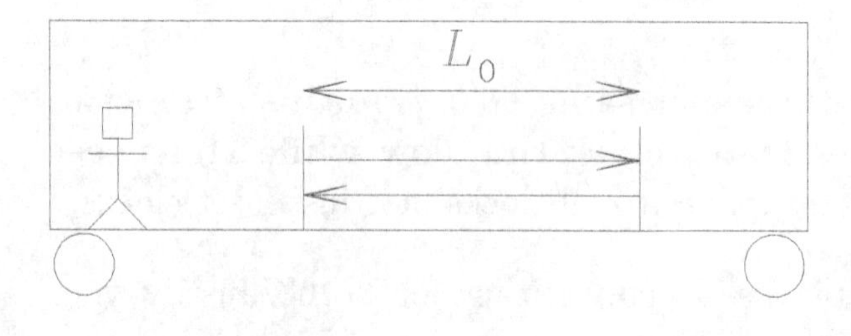

Figure 4: Hans in the same rocket train car as a horizontal laser clock of length L_0. To Hans, this clock behaves the same if the train car moves at at any **constant speed**, no matter what that speed is. A laser pulse in this clock has a round-trip time (launch to reflection to return) of $2L_0/c$.

Figure 5 below shows Hans in the same rocket train car as in Figure 4, but now moving to the right at speed v relative to Liese, who is watching from outside the train car and sees what is drawn in this figure. The horizontal laser clock is now observed to have length L_V.

4. How will L_V and L_0 be related if $v = 0$? (This is not a trick question). ________________

The train car is shown at three different times: zero (top), t_1 (middle), and t_2 (bottom). At time zero a laser pulse is launched to the right from mirror m_1 toward mirror m_2, but the train car is moving and the laser pulse has to catch up with mirror m_2 before it can reflect. It therefore travels a distance $d_1 = ct_1 = L_V + vt_1$ in time t_1 before it is reflected. On the laser pulse's return trip the mirror m_1 is approaching the pulse, so the pulse travels a distance $d_2 = ct_2 = L_V - vt_2$ in time t_2 to complete its round-trip.

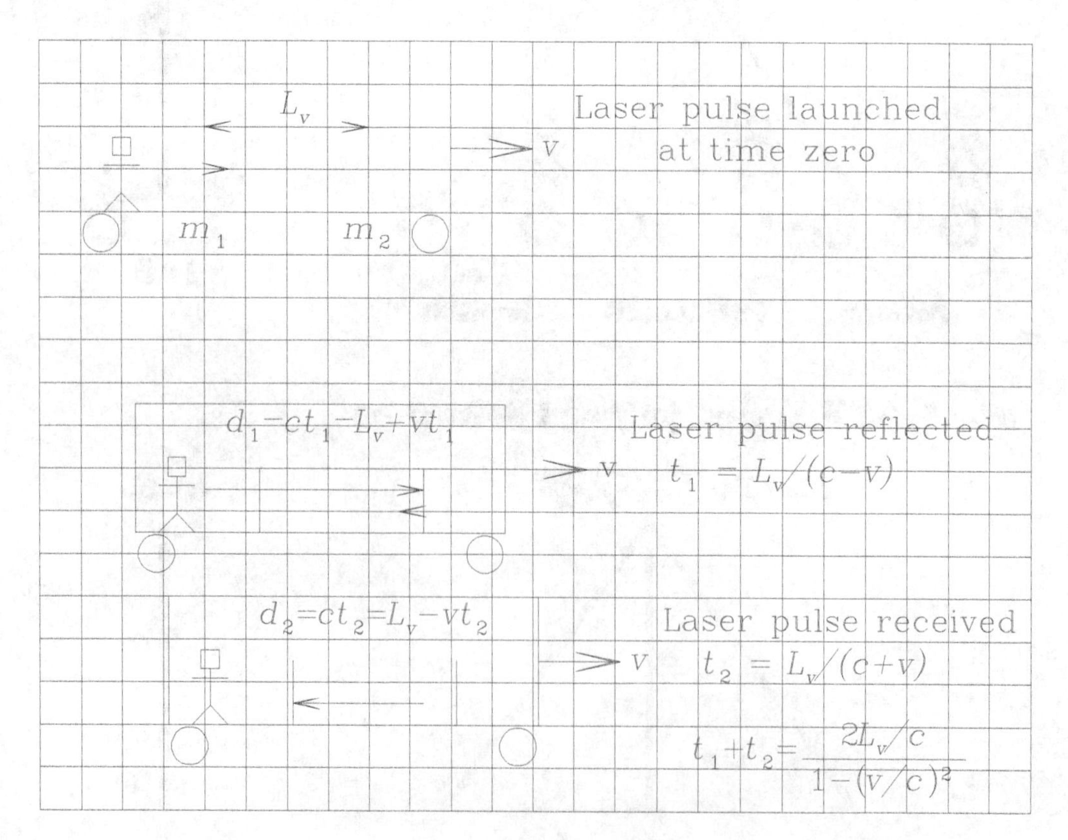

Figure 5: Hans and a horizontal laser clock in a moving rocket train car, as seen by Liese.

Therefore, to Liese the laser pulse's round-trip does not take time $2L_V$ but instead takes time $t_1 + t_2 = (d_1 + d_2)/c = (2L_V/c)/[1-(v/c)^2]$.

Inside his train car (Figure 6), Hans measures a laser light pulse round-trip time of $2L_0/c$.

Liese knows that because of time dilation, she would have seen that round-trip take a length of time $t' = (2L_0/c)/\sqrt{1-(v/c)^2}$. But Liese has also measured the round-trip time as $t_1 + t_2$.

Both of the ways that Liese can measure the laser's round-trip time must give the same answer, so she can set them equal to each other ($t_1 + t_2 = t'$) and solve for L_V in terms of L_0:

$$\frac{(2L_V/c)}{1-(v/c)^2} = \frac{(2L_0/c)}{\sqrt{1-(v/c)^2}} \quad \longrightarrow \quad \boxed{L_V = L_0\sqrt{1-(v/c)^2}}$$

Because $\sqrt{1-(v/c)^2} \leq 1$ for all v, an object measured to have length L_0 by someone who is **not** moving relative to it has a **shorter** length L_V when seen by someone who **is** moving relative to it. This effect is known as **length contraction: moving objects** (and their contents) **appear shortened along their direction of motion** (Figure 6).

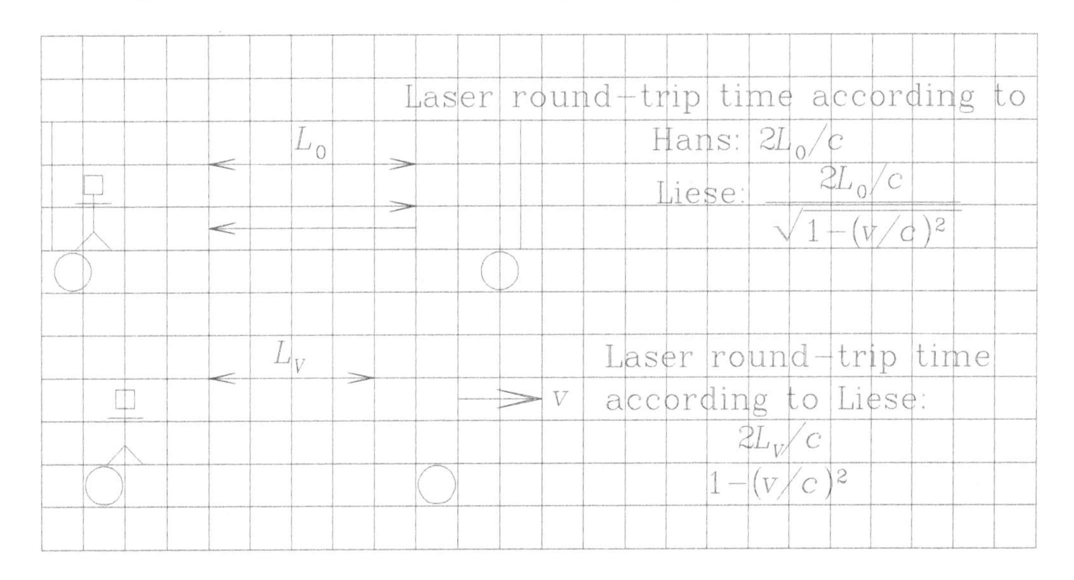

Figure 6: The same rocket train car, as seen by someone who is not moving relative to it (top) and by someone who is moving to the left with speed v (bottom), so that from their point of view the train car appears to move to the right with speed v. The length contraction effect is illustrated for that case: the moving train car appears shortened in the direction it is moving.

5. Suppose that Liese was in her own rocket train car, identical to Hans', while she watched Hans move past at a speed v relative to her. Would her train car (and everything in it) have appeared shorter to Hans? Why or why not?

Or would it have appeared normal to Hans because only Hans was moving? Why or why not?

Or would it have appeared longer because Hans' train car was shortened? Why or why not?

---

6. **(Optional)** On the back of this page, show that $t_1 + t_2 = (2L_V/c)/[1-(v/c)^2]$ (Figure 5).

Worldlines and Light Travel Time

Name:____________________ **ID#:**_______________ **Date:**_______________

Imagine that last night you (on Earth in the Milky Way galaxy) observed the brightest, bluest star in the small galaxy UGCA205 (we shall call it "your star" from now on). From its spectrum, you classified your star as **spectral type** O2. You consulted a textbook and found that an O2 star:

- has 20 times the mass of the Sun (20 **solar masses**);
- spends 2.5 million years (2.5 Myr) on the **main sequence** after it is born; then
- spends 0.5 million years (0.5 Myr) as a **supergiant** star before exploding as a **Type II supernova** and becoming brighter than the rest of the galaxy for a few weeks.

From the luminosity and temperature you measured for your star, you have determined that **the light from your star that reached Earth last night was emitted just when your star left the main sequence** (i.e., just before it started its supergiant phase).

1. How old did your star appear to you when you observed it last night? ______________

Spacetime Diagrams

The galaxy in which your star is found, UGCA205, is located approximately 5 million **light years** (Mly) from our home galaxy, the Milky Way. Because 1 light year is the distance traveled in 1 year, the light you saw from your star in UGCA205 last night left that galaxy 5 million years ago. The finite speed of light means that we always see the past and never the present.

We can illustrate the effects of the finite speed of light using a **spacetime diagram** like that of Figure 1. The horizontal axis of Figure 1 plots the distance from UGCA205 in one direction, measured in millions of light years (Mly). The vertical axis of Figure 1 plots the time since your star's birth, measured in millions of years (Myr).

On a **spacetime diagram** like the one on the next page, we can show the motion over time of objects and light through one dimension of space (motion in the other two dimensions of space is not shown on this diagram). In this diagram, at every point in time an object can be plotted at its distance from UGCA205. Each distance and time combination is a unique **spacetime location**. The line joining all those spacetime locations is called that object's **worldline**; it shows that object's position in the one space dimension covered by this diagram at all times.

By definition, UGCA205 is always located at zero on the horizontal axis (zero distance from itself). Therefore, the worldline of UGCA205 and all the stars in it is a vertical line (shown as the leftmost solid vertical arrow in Figure 1).

2. The spacetime location of the birth of your star in UGCA205 is shown by the filled triangle at coordinates (0,0). The spacetime location of your star leaving the main sequence is shown by the filled square. **Draw a dot on Figure 1** to show the spacetime location where your star explodes as a supernova.

Our Milky Way galaxy is located 5 Mly from UGCA205. Because the Milky Way is not moving significantly relative to UGCA205, the Milky Way's worldline is also a vertical line (the middle solid vertical arrow in Figure 1). Roughly 2.5 Mly from the Milky Way in the opposite direction from UGCA205 lies the Andromeda Galaxy ("Andromeda"). The worldline of Andromeda is shown as the rightmost solid vertical arrow in Figure 1.

3. How far is UGCA205 from Andromeda? (Assume that UGCA205 and Andromeda are exactly opposite each other on the sky as seen from the Milky Way.) ____________________

Looking at the Past

Light travels at a constant speed of 1 Mly per Myr (1 million light years every million years), so **light travels along diagonal worldlines** in a spacetime diagram. When your star was born, some of its light was emitted directly toward the Milky Way. The worldline of that light is shown as the right-hand dashed arrow in Figure 1. The left-hand dashed arrow shows the worldline of light emitted directly away from the Milky Way by your star when it was born.

4. Draw the worldline of light emitted toward the Milky Way from your star when it was 2.5 Myr old. How long after the star was born did that light reach the Milky Way? ______________

5. You could only see your star leaving the main sequence once the light it emitted at that time reached you here on Earth, in the Milky Way. Draw a horizontal line that includes the spacetime location of your observation of that light last night. Label it "Now".

6. How long ago was your star born? ____

7. The light from your star still exists, but does your star still exist or has it already exploded in a supernova? ____

How long until we see the light from your star's supernova? ____________________

8. An alien in Andromeda also observed your star in UGCA205 last night. How long ago did your star emit the light that reached Andromeda last night? ______________________
Based on that, how old did your star appear to the alien last night? ____________________

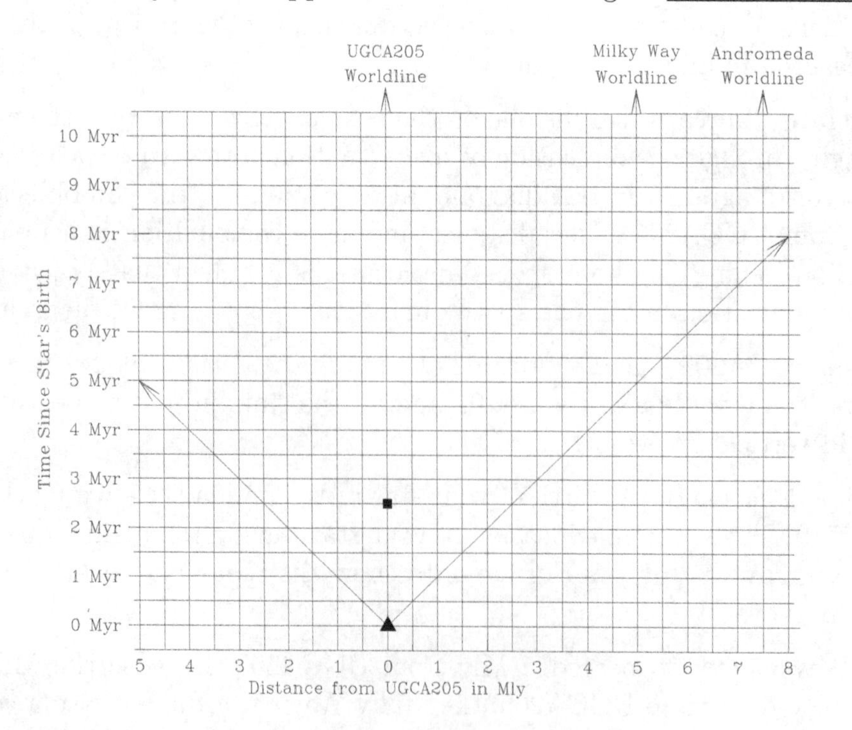

Figure 1: A spacetime diagram for UGCA205, the Milky Way, and Andromeda.

What Lurks at the Center of the Milky Way?

Name:________________________ **ID#:**________________ **Date:**________________

Beginning in the 1990s, astronomers observed stars orbiting around a radio source called Sagittarius A-star (Sgr A*) located at the center of our Milky Way galaxy (the Galactic Center). The orbital periods and orbital sizes of these stars can be used to measure the mass of Sgr A*.

Orbits Seen at an Angle

The stars orbiting Sgr A* do *not* form a disk orbiting in a single **plane** (a thin, flat region of space). They orbit in random directions and orientations. When we look at their orbits on the sky we are not looking straight down onto the planes of their orbits, but at different angles.

Consider what happens when an orbit is viewed from different angles. In the figure below, orbit A is really a circle (top left). But when viewed at an angle, it looks like an ellipse on the sky (other top row entries). To visualize this, tilt a coin at different angles. The greater the **viewing angle**, the **more eccentric the ellipse**.

In contrast, orbit B is really an ellipse (bottom left). When viewed from the right direction at different angles, it can appear as a circle or an ellipse on the sky (other bottom row entries).

For all orbits, in the figure below the filled dot shows the **true foci** while the open dot shows the **apparent foci** on the sky at that viewing angle.

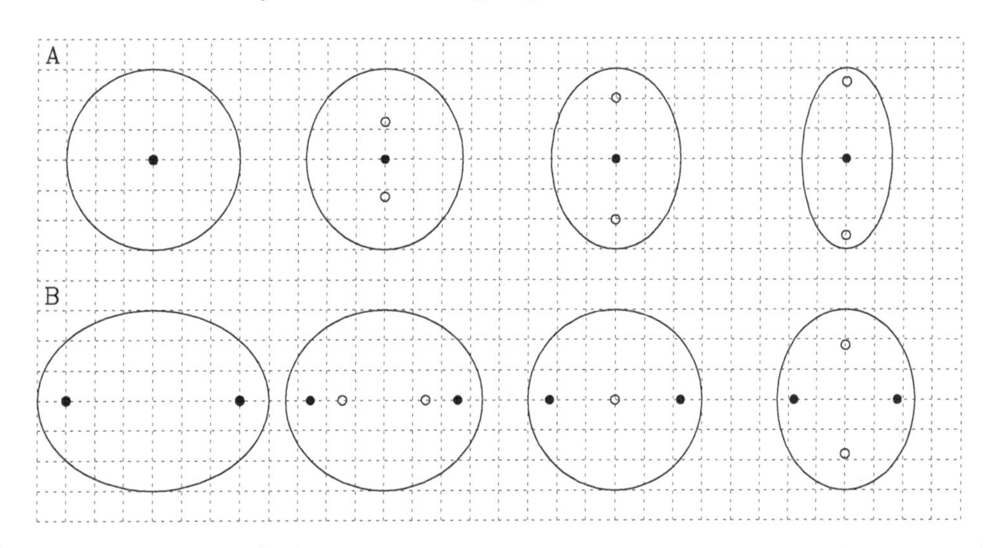

1. The **apparent eccentricity** of an ellipse, e_{app}, is the distance between its apparent foci divided by the apparent length of its major axis. When an orbit is viewed from an angle other than straight down, are there angles where e_{app} increases? *(Always | Sometimes | Never)* Decreases? *(Always | Sometimes | Never)* Stays the same? *(Always | Sometimes | Never)*

2. The **apparent semi-major axis** of an ellipse, a_{app}, is half the length of its apparent major axis. When an orbit is viewed from an angle other than straight down, are there angles where a_{app} increases? *(Always | Sometimes | Never)* Decreases? *(Always | Sometimes | Never)* Stays the same? *(Always | Sometimes | Never)* Therefore, can an ellipse ever have an apparent semi-major axis on the sky a_{app} larger than its true semi-major axis a? *(Yes | No)*

From Angular Sizes to Physical Sizes

The stars around Sgr A* have orbits whose apparent semi-major axes are measured in **angular sizes** on the sky. To convert to physical sizes in **astronomical units (AU)** requires knowing the distance of Sgr A*. A **kiloparsec** (about 3300 light-years) is convenient for distance measurements because an orbit 1000 AU in radius around an object 1 kiloparsec away will have a radius on the sky of 1 arcsecond. Therefore, **an orbit 1 AU in radius around an object 1 kiloparsec away will have a radius on the sky of 1 milliarcsecond (mas).**

3. Sgr A* is located 8 kiloparsecs away. If you measure an orbit around Sgr A* to have a semi-major axis of 1 mas, what is the semi-major axis of that orbit in AU?

____

4. Table 1 lists the semi-major axes in mas of four stars orbiting Sgr A*. Use the conversion factor from the previous question to convert those measurements to semi-major axes in AU.

A Limit on the Mass of Sgr A*

Newton's version of Kepler's Third Law tells us that the semi-major axis (a) and **orbital period** (P) of anything orbiting a much more massive object of mass (m) are related by:

$$\frac{a^3}{P^2} = m \quad \text{(for } a \text{ in AU, } P \text{ in Earth years, and } m \text{ in solar masses)}$$

This relationship does not depend on the eccentricity of the orbit, only on its semi-major axis. We can therefore use stars orbiting Sgr A* to get an **apparent value for the mass** of Sgr A*, m_{app}, using the stars' P and apparent semi-major axis a_{app}. Because a_{app} is never larger than the true semi-major axis a, m_{app} will never be larger than the true mass m; therefore, we know that m_{app} is a **lower limit** to m. Each star will give a different value of m_{app}, and the true mass of Sgr A* will be at least as large as the largest of those values.

5. Calculate and enter the value of m_{app} for each star in Table 1.

6. When the orbits of these stars are modelled in three dimensions, a mass of 4 million solar masses is found for Sgr A*. Are the values of m_{app} you found consistent with this value of m? *(Yes | No)* Despite its large mass, the **luminosity** of Sgr A* is less than 40 times the **luminosity** of the Sun. An object with 4 million times the mass of the Sun but less than 40 times the luminosity of the Sun is most likely a ____________________.

Table 1: Apparent orbital data for stars near Sgr A*. Data adopted from Ghez, A., et al. 2005, The Astrophysical Journal, 620:744, "Stellar Orbits around the Galactic Center Black Hole".

Star Name:	S0-2	S0-16	S0-19	S0-20
Period in Earth years	15	36	37	43
Apparent Semi-Major Axis in mas	95	195	200	220
Apparent Semi-Major Axis in AU				
Lower Limit to Sgr A* Mass (m_{app})				

Quasar Jets: a Superluminal Optical Illusion

Name:__________________________ **ID#:**____________________ **Date:**____________________

At the center of every massive galaxy lies a supermassive black hole. In the distant past, each black hole was surrounded by a vast gas accretion disk that reached very high temperatures and outshone the galaxy's stars. Such an object is known as a **quasar**. These accretion disks can also launch narrow **jets** of matter consisting of blobs of gas moving at nearly the speed of light. Here we'll learn how these jets can *appear* to be superluminal (faster than light).

Figure 1 (next page) shows the path of a blob of gas ejected from a quasar at the upper left (labeled Q, at coordinates 0, 0). The blob is ejected on a day we'll call day 0.

Earth is located towards the bottom of the page. We know the distance from this quasar to Earth (vertical scale), so we know what angle on the sky equals one light-day at the quasar's distance (1 light-day is $\frac{1}{365}$ light-year). Those distances are shown on the horizontal scale.

The blob is constantly emitting radio waves towards the Earth and in other directions as well. Those radio waves travel at the speed of light c: 1 light-day per day. The blob moves with speed $v = \frac{5}{6}c$ in a slightly different direction than the direction towards Earth. The positions of the blob on day 6 and day 12 have been marked for you.

Let's investigate how fast the blob appears to move across the sky as seen from Earth.

1. On day 6, the blob is located at position B and the light the blob emitted towards Earth on day 0 is located at position (fill in the blank) ____________________________________.

2. On day 12, at what position is the light the blob emitted towards Earth on day 0 located? __________________________________ On day 12, at what position is the light the blob emitted towards Earth on day 6 located? __________________________

The dashed vertical lines show the paths taken to Earth by light emitted by the blob on days 0, 6, and 12. Earth is located 10 billion and 20 light-days from the quasar. We have removed 10 billion (10^9) of those light-days, and only show the last few light-days before the light reaches Earth. When light emitted by the blob at some position on the sky crosses the solid horizontal line at the bottom representing Earth, the blob is seen at that position in Earth's sky.

3. On day 14, the light the blob emitted on day 0 is located at N, the light from day 6 is at L, and the light from day 12 is at J. Connect those points and imagine that light continuing to move towards Earth. On day $10^9 + 20$, the light the blob emitted on day 0 is located at X, and the light from day 6 is located at ________. The light emitted by the blob on day 0 reaches Earth on day $10^9 + 20$, so when does the light emitted by the blob on day 6 reach Earth?

____

4. How many days pass on Earth between the arrival of light emitted on days 0 and 6? ______ How far will the blob appear to move on the sky in that time? __________________________

5. We know the blob is actually travelling at speed $v = \frac{5}{6}c$, but we can define *apparent speed* $v' =$ *(apparent distance moved on sky)* $\div$ *(apparent time needed to move that distance)*. What is the apparent speed of the blob across the sky? ____________________________________

6. This blob almost catches up to the light it emits, making it appear to move faster than light. Do you think you'd measure an apparent speed $v' > c$ for a blob that takes twice as long to reach point B (12 days instead of 6)? ____________ What would v' be in that case? ____________

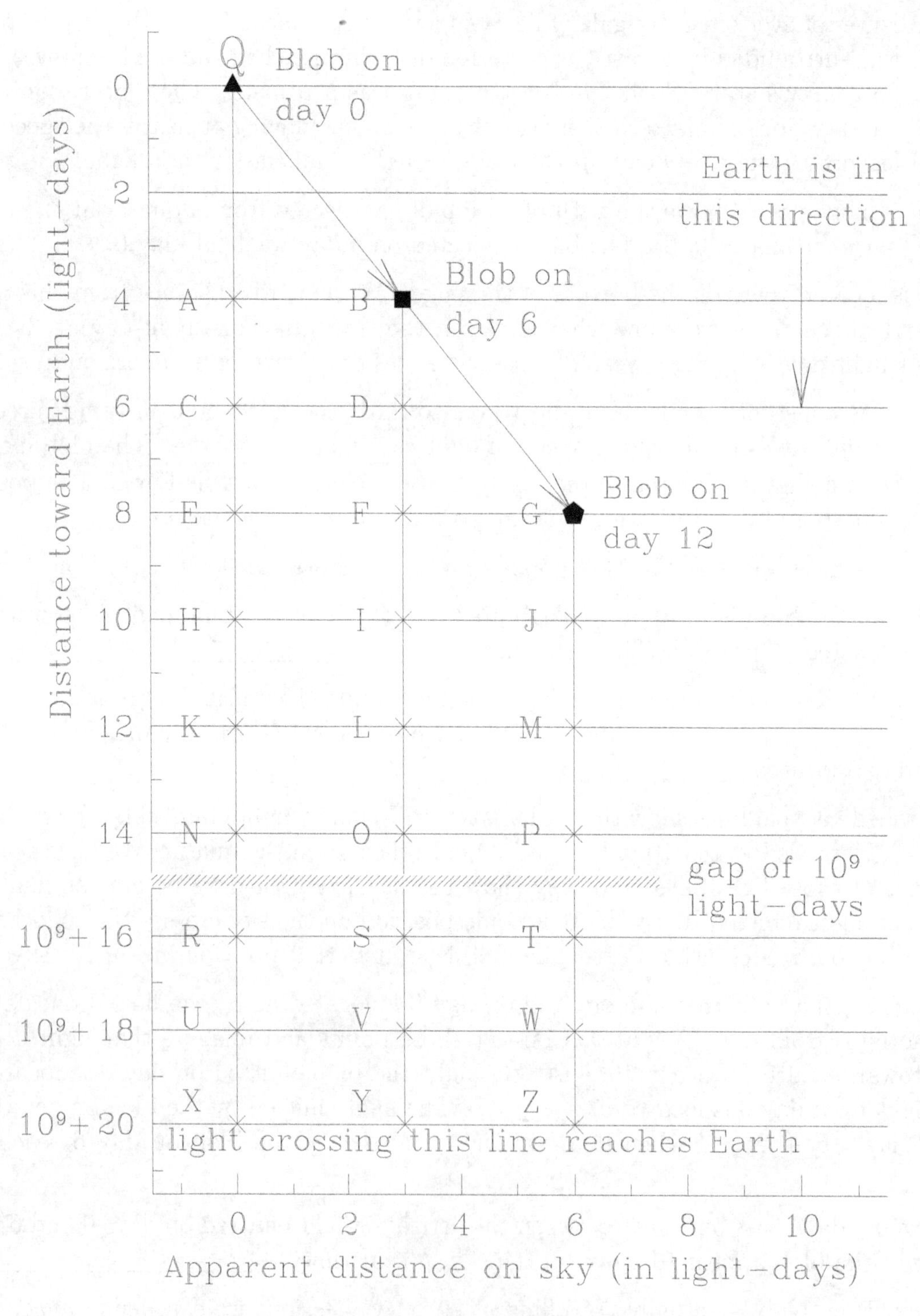

Figure 1: A blob of gas ejected with speed $v = \frac{5}{6}c$ from a quasar (located at 0, 0) and emitting light towards the Earth. When light crosses the solid line at the bottom of the diagram, 10,000,000,020 light-days away from the quasar, that light reaches the Earth.

The Expansion of Space

Name:________________________ **ID#:**_________________ **Date:**_________________

In the first half of the 20th century, astronomers, including Edwin Hubble, studied the velocities of galaxies near our Milky Way galaxy for the first time. In every direction they looked, they found that distant galaxies had velocities directed away from the Milky Way. From this they concluded that **space itself is expanding**, carrying all galaxies with it.

It is difficult to illustrate the expansion of three-dimensional space on a two-dimensional page or screen. Instead, consider a universe with two space dimensions on the surface of a balloon that expands in time. Creatures in such a universe can only move in two dimensions on the surface of the balloon, just like you can only move North-South and East-West on the surface of the Earth. With every instant of time that passes, the distance between any two points on the surface of the balloon increases. From the point of view of the two-dimensional creatures living in two-dimensional galaxies on the balloon surface, space itself has expanded.

Figure 1 shows a top view of one hemisphere of a two-dimensional "balloon" universe, projected down onto the page at three different points in time from two different perspectives: centered on galaxy **a** in the left column, and centered on galaxy **b** in the right column. The axes to the right of each image give that image's distance scale in millions of light-years (Mly).

1. Enter in the table the time elapsed between each image of the universe shown in Figure 1, measured in units of gigayears: 1 gigayear (1 Gyr) = 1 billion years.

2. Enter the distance between galaxy **a** and **b** on each image.

3. Enter the distance between galaxy **a** and **c** (or **b** and **c**) on each image. Complete the following sentence: Galaxy **c** is ________________________ as far away from galaxy **a** as galaxy **b** is. (If you measured from galaxy **b**, switch **a** and **b** in the above sentence.)

4. Calculate the **velocity** of galaxy **a** relative to galaxy **b**: the change in the distance between them since the last image divided by the time since the last image. Then do the same for galaxy **a** (or **b**) relative to **c**. Complete the following sentence: The velocity of galaxy **c** relative to galaxy **a** is ________________________ as large as the velocity of galaxy **b** relative to galaxy **a**. (Again, if you measured from galaxy **b**, switch **a** and **b** in the above sentence.)

5. Based on your results, complete the following sentence: The farther apart two galaxies are in an expanding universe, the ________________________ the **relative velocity** between them.

Image	Age of Universe (Gyr)	Time Since Last Image (Gyr)	Distance, **a** to **b** (Mly)	Velocity of **a** relative to **b** (Mly/Gyr)	Distance, (**a** or **b**) to **c** (Mly)	Velocity of (**a** or **b**) relative to **c** (Mly/Gyr)
Top	2.5 Gyr	—	25 Mly	—		—
Middle	5.0 Gyr					
Bottom	7.5 Gyr					

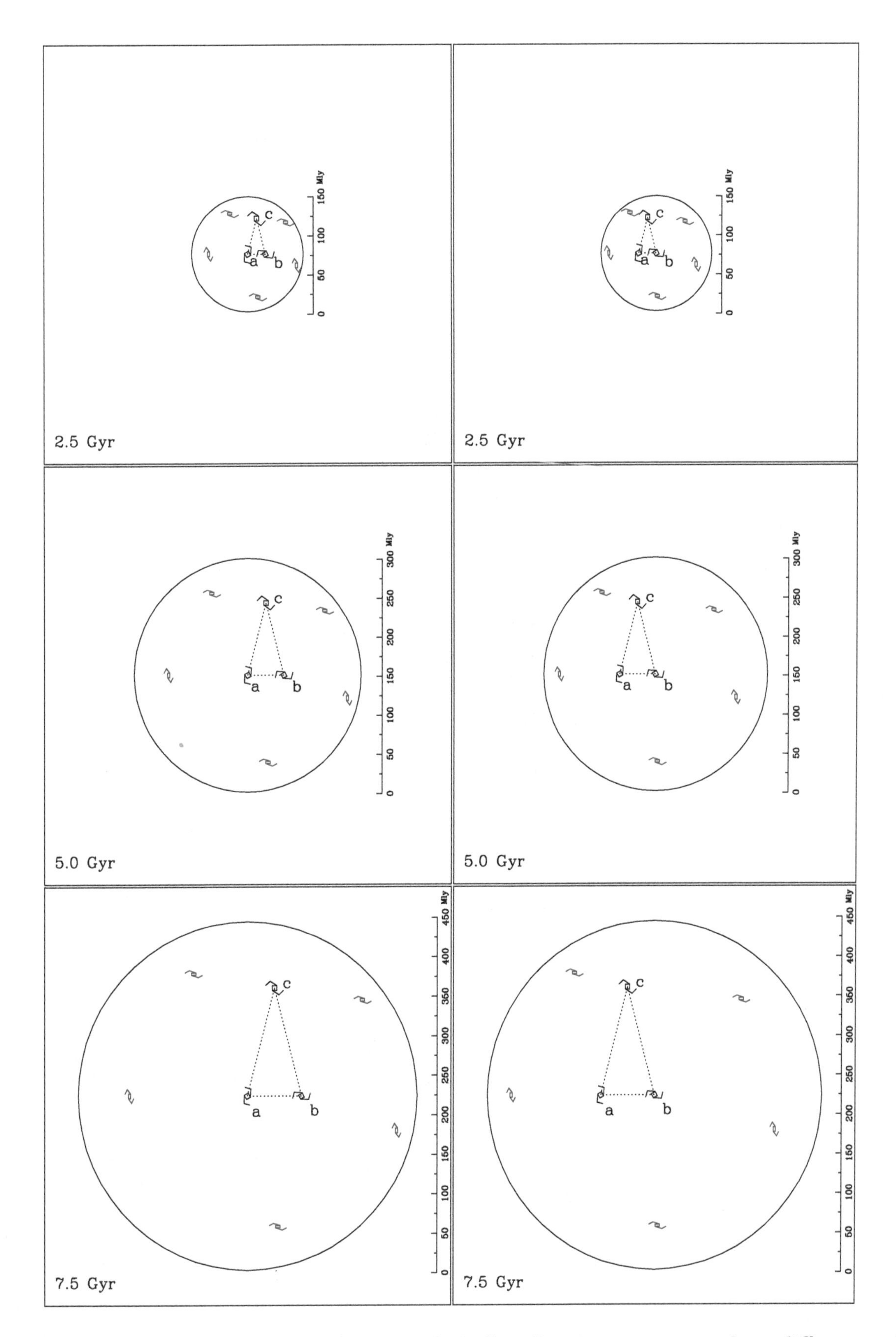

Figure 1: One hemisphere of a two-dimensional "balloon" universe, seen at three different points in time (every 2.5 billion years or gigayears (Gyr), top to bottom) from two different perspectives: centered on galaxy **a** (left-hand panels) and centered on galaxy **b** (right-hand panels).

The Acceleration of the Milky Way

Name:________________________ **ID#:**_________________ **Date:**_________________

Figure 1: A map of the microwave emission of the entire sky, projected onto a two-dimensional map. Lighter colors mean a higher temperature; darker colors mean a lower temperature.

From Doppler Shift to Velocity to Acceleration

Figure 1 shows a map of the sky made using microwave light. In this map, everywhere across the sky we can see the **cosmic microwave background radiation** (CMBR). The temperature of the CMBR is almost equal in all directions, but this map shows the small variations in temperature caused by the motion of the Sun relative to the CMBR (the Doppler effect). Lighter colors mean a higher temperature; darker colors mean a lower temperature.

1. If you are not moving relative to the CMBR, you will observe the peak wavelength of the CMBR to be the same in all directions. If you are moving relative to the cosmic microwave background radiation and you observe the CMBR in the direction of your motion, does the Doppler effect shift the peak wavelength of that radiation to shorter or longer wavelengths when you look in the direction you are moving? (Hint: The effect would be the same if that part of the sky was moving towards you, instead of you moving towards it.)

____

How does the peak wavelength change when you look back in the direction you came from?

____

The Doppler shift changes the peak wavelength of the CMBR across the sky, which is the same as changing the temperature of the CMBR across the sky. The amount by which the peak wavelength changes is larger for faster speeds. By measuring the wavelength change, we can figure out how fast the Milky Way is moving relative to the CMBR. (You can assume we've already accounted for the Earth's motion within the Milky Way.)

2. The percentage amplitude of the change in the peak wavelength of the CMBR across the sky equals the speed of the Milky Way as a percentage of the speed of light c ($c = 300{,}000$ kilometers per second, abbreviated as km/second). If the peak wavelength of the CMBR is 0.185% shorter in some direction, what is the Milky Way's velocity in that direction?

We believe that the Milky Way was born with zero velocity relative to the CMBR. We also know that velocity is acceleration multiplied by time. We can therefore use the age of the universe and the Milky Way's current velocity to determine the acceleration of the Milky Way.

3. If the Milky Way is 13.75 billion years old (13,750,000,000 years) and is now travelling at 555 km/second, how many centimeters per second faster is the Milky Way moving every 1000 years? (Hint: 1 km = 100,000 cm.)

That number is the Milky Way's acceleration, in centimeters per second per 1000 years.

The Milky Way IS moving at about 555 km/second towards a point in the constellation of Hydra (the water snake), near the constellation of Crater (the cup). Close to that point on the sky, between Hydra and the constellation of Centaurus the Centaur, lies a supercluster of galaxies known as the Shapley Concentration. This supercluster is located about 600 million light-years (Mly) away from the Milky Way. (For comparison, the Andromeda galaxy is located 2 million light-years away.) Could this supercluster be responsible for accelerating the Milky Way? Knowing the distance d to that supercluster, we can use our estimate of the Milky Way's acceleration a to estimate the supercluster's mass M, since Newton's Law of Gravity tells us that

$$a = \frac{GM}{d^2} = \frac{GM_{gal}N_{gal}}{d^2}$$

where N_{gal} is the number of galaxies in the supercluster and M_{gal} the average mass of a galaxy.

4. When a is measured in (cm/second)/1000 years, M_{gal} is measured in Milky Way masses, and d is measured in Mly (million light-years), the above formula becomes $a = 3N_{gal}/d^2$. Using that formula, if a supercluster of galaxies located $d = 600$ Mly away is responsible for accelerating the Milky Way to its observed velocity, what number N_{gal} of galaxies like the Milky Way would be in it?

---

5. Optical and X-ray images of the Shapley Concentration show it consists of about 18 rich galaxy clusters, and that each rich galaxy cluster contains the mass of 8000 galaxies like the Milky Way. Given the total mass of the galaxy clusters in the Shapley Concentration, what fraction of the Milky Way's acceleration can it explain?

Civilizations in Our Galaxy: the Drake Equation

Name:________________________ **ID#:**________________ **Date:**________________

Let's estimate step-by-step how many technological civilizations with whom we might communicate might exist in our galaxy. Astronomer Frank Drake made the first estimate of this kind in 1961, and the equation he used to make the estimate is named after him.

There are different ways to write the Drake equation, because there are a huge number of factors that could affect this final result. We will try to focus on the most important ones, and the ones for which we can obtain reasonable estimates. Our version of the Drake equation is:

$$N_{\text{civilizations}} = \frac{N_{\text{new stars}}}{\text{year}} \times f_{\text{planets}} \times \frac{N_{\text{habitable planets}}}{\text{star}} \times f_{\text{complex life}} \times f_{\text{technological}} \times \text{Lifetime}_{\text{civilization}}$$

where each f value is a fraction less than one and each entry is one step in the calculation:

- $N_{\text{new stars}}/\text{year}$ is the number of new stars born in the Milky Way each year;
- f_{planets} is the fraction of those stars with planets;
- $N_{\text{planets}}/\text{star}$ is the average number of habitable planets per star **with planets**;
- $f_{\text{complex life}}$ is the fraction of planets that develop complex life;
- $f_{\text{technological}}$ is the fraction of planets with complex life that develop a technological civilization;
- $\text{Lifetime}_{\text{civilization}}$ is the averge lifetime of a technological civilization, in years.

1. Use the Drake equation to make your own calculation of $N_{\text{civilizations}}$, the current number of technological civilizations in the Milky Way. Questions 2 to 7 are each one step towards the answer. You can decide to be optimistic or pessimistic at each step, or you can make an optimistic and a pessimistic assumption for each step to see how much your final estimates differ depending on the assumptions you make. Write your answers from Questions 2 to 7 below:

$N_{\text{civilizations}} =$ ________ × ________ × ________ × ________ × ________ × ________ = ________

Multiply those numbers together and write the result after the equal sign above. Based on that answer, do you think it is likely that humans will one day detect evidence of an alien civilization?

---

2. $\mathbf{N_{\text{new stars}}/\text{year}}$: Let's assume that civilizations only develop on planets around stars. How many stars are born in our galaxy each year? (Hint: The disk of our galaxy has about 100 billion stars and is about 10 billion years old.)

3. $\mathbf{f_{\text{planets}}}$: What fraction of those stars have planets? (Hint: Radial velocity extra-solar planet searches find planets around 30% of stars. What does that imply about the true fraction?)

4. **$N_{\text{habitable planets}}$/star** : A habitable planet is one that has a gas or liquid atmosphere and a temperature range suitable for simple life. How many *habitable* planets (or moons) are there per star with planets, on average? Note that this number can be less than one. For example, a value of 1/10 would mean one habitable planet for every ten stars with planets.

Based on our solar system only, what is the possible range of values for this quantity?

Most stars are less luminous than the Sun, meaning any habitable planets around them must be closer to the star than if they orbited the Sun. Will that requirement make the number of habitable planets per star smaller or larger than estimated above?

So what is your final estimate of $N_{\text{habitable planets}}$/star?

5. **$f_{\text{complex life}}$** : Let's assume that all habitable planets (and moons) develop simple life immediately after formation. What fraction of planets develop complex life (animal life that could potentially reach human-level intelligence)? (Hint: On Earth complex life took almost 4 billion years to develop, and the Sun only lives for about 10 billion years.)

6. **$f_{\text{technological}}$** : What fraction of planets with complex life develop technological civilizations able to communicate through space?

7. **$\text{Lifetime}_{\text{civilization}}$** : For how many years do technological civilizations survive? Let's call a civilization "technological" if it has developed radio transmitters and telescopes so that it can send and receive radio signals to other civilizations. (That definition gives a lower limit on how long technological civilizations last, using Earth as our only example.)

You have now calculated all the values you need to for the Drake equation in Question 1, so turn back a page and plug in to find $N_{\text{civilizations}}$.

8. The Drake equation estimates how many civilizations might exist in our galaxy with whom we might *communicate*. What other factors would we need to consider if we wanted to estimate how far we would have to travel to *meet* with one of those civilizations directly?